WILD
PLACES
OF GREATER
BRISBANE

To Stephanie
with love Christmas 1996.

Ian + Gay.

WILD PLACES

OF GREATER

BRISBANE

Stephen Poole
& others

museum

Published by the Queensland Museum

with the support of the Environment Management Branch
of Brisbane City Council

Brisbane City

and the assistance of the Department of Natural Resources
and the Department of Environment

QUEENSLAND GOVERNMENT
Department of Environment

**NATURAL
RESOURCES**

museum

Requests for this book should be made to:

Queensland Museum
PO Box 3300
SOUTH BRISBANE Q 4101
Australia
Fax: (07) 3846 1918
International Fax: +617 3846 1918

National Library of Australia
Cataloguing-in-Publication data:
Wild places of Greater Brisbane
Includes index.
ISBN 0 7242 7110 4

1. National parks and reserves – Queensland – Brisbane Region.
2. Forest reserves – Queensland – Brisbane Region.
3. Brisbane Region (Qld.) – Guidebooks.
I. Poole, Stephen, 1967– . Queensland Museum.
919.431

Edited by Michelle Ryan
Designed by Diane Yeo and Andrew Ness
Photography by Bruce Cowell
Illustrated by Robert Allen and Sally Elmer

With thanks to Pat Comben, Greg Czechura, Derek Griffin, Tim Moore and Greg Oliver.

Set in 8.5 pt Utopia
Printed on Silk Matt by Prestige Litho, Brisbane

Published by the Queensland Museum with the support of the Environment Management Branch of Brisbane City Council.

First Printed 1996.
Photographs on pp. 22, 27, 29, 33, 46, 53, 58, 59, 61, 62, 65, 83, 87, 92, 97, 110, 114, 124, 128, 160, 164, 168, 171, 172, 200, 202-206.
© Bruce Cowell.

FOREWORD

Late in 1995 the Queensland Museum published *Wildlife of Greater Brisbane*. The phenomenal success of this book reinforced what we at the Museum have long believed — that the general community is vitally interested in the natural environment.

It is with this in mind that the Museum decided to publish this companion volume, *Wild Places of Greater Brisbane*. The first book introduced readers to some of the wonderful animals that live in the Greater Brisbane Region; this new book tells readers where the animals might be seen. But more than this, *Wild Places* reveals some of the superb, and often surprising, natural habitats that can be found within a 180 km radius of city.

As natural history specialists, museum scientists are only too aware of the fragility of our environment. To ensure the survival of animal (and plant) species, their habitats must be preserved. Every piece of bushland that is lost exerts new pressure on the remaining sanctuaries and it is not only rare and endangered species that bear the impact. In time, animals and plants that are now considered familiar and common may also diminish.

The conservation of existing habitats is important but where natural environments have been degraded or lost it is up to all of us to take positive steps to heal the landscape. In doing so we will ensure that these areas — and the animals and plants that are part of them — will be safeguarded for generations to come.

This book was a collaborative project and would not have been possible without the generous support of the following organisations:

The Environment Management Branch of Brisbane City Council contributed financially to the project and provided a substantial portion of the text. Brisbane Forest Park and the Department of Natural Resources (Forestry) assisted with text, as did the Department of Environment. The staff of all these agencies are to be commended for their bipartisan support. The John Oxley Library provided the historical photographs on pages 1-6.

Many individuals deserve thanks, in particular the authors — Adrian Caneris, Pat Comben, Greg Czechura, Shawn Delaney, Martin Fingland, Toni Hess, Rod Hobson, Peter J. M. Johnston,

Tim Low, David Morgans, Mark Peacock and Stephen Poole. Thanks also to Steve Finlayson (Main Range National Park), Brett Waring (Department of Natural Resources), Andrew Greenall (Department of Environment), Tim Moore (Department of Environment) and Bill McDonald (Queensland Herbarium).

Alan Bartholomai

CONTENTS

A MESSAGE FROM DAVID BELLAMY

I believe that letting people experience our natural world is the best form of environmental education. In my previous visits to Brisbane, the tools to achieve this were sadly lacking, but in less than a year, two projects have changed this.

Firstly, *Wildlife of Greater Brisbane* enabled residents to discover their local animals. Now, this companion volume, *Wild Places of Greater Brisbane* encourages people to explore some of the wonderful natural areas where these animals live. These books are among the best ways to encourage people to learn about and come to appreciate their environment.

The Greater Brisbane Region is fortunate to have so many magnificent natural assets close by — rainforests, eucalypt bushlands, wetlands and the barrier sand islands of Moreton Bay. But imagine the region just over two centuries ago — a mere blink in time — before the trappings of modern civilisation altered it forever. The surviving natural habitats have stood watch as the land was progressively cleared for logging, agriculture and settlement — processes that have accelerated rather than diminished through the years.

To make the most of the areas that remain and their associated bio-diversity, it is up to us to ensure that they are buffered and linked to afford maximum resilience from the pressures of modern society. Loss of our wild places is a cumulative process. Imagine a spider's web, cut one supporting strand and the web still functions; cut a few more and the web becomes less effective, cut all the strands and the web collapses. So it is with our natural areas. We should strive to covet them, to keep them as close to the original whole as possible — to repair and extend, not denude and fragment. After all, who knows what harm will befall us if we lose everything that was?

Take this book, tell your friends and visit at least one of the wild places herein. See the plants and animals — and reflect on what we have, what we had and what the future might hold. The future is ours to shape and to determine. Let us come back from the brink and mould a sustainable future for all of us.

Professor David Bellamy
THE CONSERVATION FOUNDATION, September 1996

HOW TO USE THIS BOOK

The term "wild places" has different meanings for different people. For many, a "wild place" is anywhere outside the structured familiarity of our suburbs. For others, the only true "wild places" are remote and isolated from all visible trace of human society.

Between these two extremes is a range of experience and expectation. A popular swimming hole in a national park, well-worn tracks through regenerating bushland or an inaccessible rainforest gorge — all are "wild places".

The aim of this book is to raise community awareness about the "wild places" of the Greater Brisbane Region and perhaps, in the long-term, encourage community appreciation of them.

It is not necessary to be a hardy bushwalker to discover some truly wonderful parks and reserves or to travel hundreds of kilometres. Most are close by and each offers a unique opportunity to explore our natural world.

This book is intended as a broad overview for general readers, rather than people with specialist knowledge or requirements. It is not meant to be a detailed bushwalking guide or a comprehensive environmental study.

The first part of the book describes the diverse natural habitats found within the region and their main characteristics as an aid to identification. The second section is a guide to more than 30 "wild places", including national parks, state forests and conservation reserves, within a 180 km radius of Brisbane.

The sites are listed in alphabetical order (with the exception of Brisbane Forest Park, one of the largest and most significant reserves), and grouped according to their distance from the Brisbane General Post Office.

Each entry has brief information on the vegetation, fauna and special features of the area. Distance, driving time, access, facilities, permits and restrictions are included in the Location Guides. However readers should consult the relevant contacts for more detailed information (listed in Useful Contacts at the end of this book).

In a small book like this, it is not possible to include every site and selection has been made on the basis of habitat quality, significance of the site and visitation levels. The Greater Brisbane Region is usually defined as the metropolitan area and a further radius of 60-100 km. However, for the purposes of this book, some parks and reserves within a 150 km have also been included. For example, Main Range and Lamington National Parks are two of the most important natural habitats in South-east Queensland and Lamington is one of the most heavily visited.

The "wild places" of the Greater Brisbane Region belong to everyone. There are a few simple guidelines which will help make your visit more enjoyable.

• Remember, all plants, animals and natural and cultural features are protected in national parks and state forests. Do not disturb anything. Do not pick flowers or take samples of plants. Try not to trample or destroy vegetation when selecting camp sites and erecting tents, or while walking.

• If you are planning to camp in a national park or state forest, contact the rangers beforehand to ensure you have the relevant information and permits.

• If you are planning to explore remote areas, always let the rangers know of your plans and timetable. Take a map and a compass and ensure your food, clothing and equipment is adequate.

• Leave domestic pets at home as they can harm native wildlife. Dogs are allowed in some State Forest areas, but check with the rangers first.

• Do not light open fires. Use fireplaces and wood where they are provided or bring your own fuel stove. Extinguish fires when leaving a site. Observe fire bans. Do not collect firewood from surrounding forests.

• Feeding native animals can be harmful for them and dangerous for humans — you could be bitten or scratched.

• Use rubbish bins where they are provided or take your rubbish home. Do not bury rubbish as native animals can uncover it.

• Use toilet facilities where they are provided. If there are no toilets, bury human wastes well away from watercourses and walking tracks.

• Protect waterways and aquatic life. Soaps, detergents, toothpaste and sunscreens pollute creeks and streams.

• When walking, stay on the tracks and follow the signs.

• Leave no trace of your visit. Make sure your campsite or picnic area is as good as, or better, than when you found it.

• Four-wheel-drive enthusiasts should ensure they have the relevant permits and should stay on constructed tracks. Do not drive "off-road".

Observing these points, and treating our "wild places" with care and respect will help preserve them now and for the future.

Logging at Springbrook, 1906 John Oxley Library

BEYOND THE BACKYARD

You and I live somewhere truly special — a biological junction between north and south, a sub-tropical paradise that many around the world would envy. And it is a paradise when compared to other cities which have lost almost all their natural surroundings. Where they survive, these areas have often been severely altered or degraded.

The Greater Brisbane Region has been more fortunate. Within the city perimeters and within a relatively short distance,[1] there are still "wild places" — rugged mountains, dense sub-tropical rainforests, quintessential eucalypt forests and idyllic coastlines. Luckily for us, the "bush", in all its forms, is not hundreds of kilometres away — but just beyond the backyard, waiting to be discovered.

Giant Kauri, 1912 JOL

The Brisbane region is the gateway to Queensland but many visitors (and locals) fail to pause and experience its natural attractions. Most people have icons fixed in their minds — the Great Barrier Reef, the Wet Tropics, Fraser Island and the Gold Coast beaches. But the Greater Brisbane Region has its own marvels — including the World Heritage listed areas of the Scenic Rim (Lamington, Springbrook, Main Range and Mt Barney), wetlands of international significance, and the sand islands of Moreton Bay — to name but a few.

The region's natural habitats include the mountainous green backdrops of the Great Dividing Range and the nearer McPherson, D'Aguilar, Darlington and Blackall Ranges. It is in these ranges that most of the region's sub-tropical and temperate rainforests occur, as do most of the wet and much of the dry eucalypt forests.

Closer to the coast are the remaining dry eucalypt forests and woodlands;

the important, but rapidly disappearing paperbark forests; and diminishing heathlands. (It is this coastal belt which is under most threat in the Greater Brisbane Region.) Along the coast there are salt marshes, mangroves and tidal mudflats — important to local fisheries and the aquatic ecosystems of Moreton Bay. The Bay sand islands — Bribie, Moreton and North and South Stradbroke — are strongholds for some plant and animal species which are threatened or already extinct on the mainland.

Brisbane has more animal and plant species than any other capital city and could truly be called the "Wildlife Capital of Australia". This is due primarily to the range of different habitats across the Greater Brisbane Region.

As well, the region falls within what is known as the Macleay-McPherson Overlap. This is the area where the northern (Torresian) and southern (Bassian) distribution limits of many plants and animals merge.

Bush picnic, Mt Tamborine, 1920 JOL

There is considerable debate about the extent of this phenomenon; however it is generally considered that the Greater Brisbane Region falls within the centre of the Overlap.

Despite this richness of fauna and vegetation, it is clear from maps of South-east Queensland, that our wild places are not well protected. Development has been the driving force for decades, mostly at the expense of natural habitats.

Within a 250 km radius of Sydney there are more than 2 million hectares of national parks, state forest, water catchment and other reserves which protect such outstanding sites as the Blue Mountains, Wollemi, Kuringai Chase, Royal, Brisbane Waters, Yengo and Kanangra Boyd National Parks.

These areas provide a barrier to the city's expansion and because of their size will, over time, maintain many of the plants and animals that exist there. (Outside the reserve system, the future is uncertain. The Hawkesbury-Nepean catchment, for example, is suffering from immense urban pressures.)

By contrast, the Brisbane region protects less than 10 percent of the Sydney reserves — within a similar 250 km radius. Unprotected natural areas held as freehold, leasehold or Crown Land are under acute threat of clearance. As well, many plants and animals are hanging on in habitats that are too small to ensure their long-term viability.

This situation has evolved only during the past 170 years. The first Europeans saw an almost pristine, awe-inspiring wilderness. In 1825, one convict remarked, "It looked as though some race of men had been here before us and planted this veritable Garden of Eden".[2]

But it was an Eden which few Europeans ever appreciated. Most considered it an unlimited resource for materials and land. The unfamiliar beauty of the rainforests was lost on them, perhaps because beauty could not be harvested.

The Aboriginal people were at home in the place that today is known as the Greater Brisbane Region. They used the forests and the animals within them as sources of food and material products. The sticky sap of Strangler Figs was used to trap small birds.[3]

Kauri Pine, Queensland.

Twine baskets were fashioned from parts of the Cabbage Tree Palm and certain eucalypts. The bark of melaleuca trees was used as roofing for shelters. Of the native plants found in the region, Aboriginal people used more than 200 species for food and medical treatments.[4]

Their use of the forests was sustainable — in contrast to the European settlers who relentlessly cleared the bush for the seemingly endless supplies of giant Red Cedar, Yellow Carabeen and Hoop Pine, and then for agriculture (mostly pasture) and settlement.

This pattern was repeated in adjacent regions as with the loss of the "Big Scrub" rainforest in far northern New South Wales which diminished from an estimated 75,000 ha in 1874 to less than 1 percent in 1900. (Sustainability is a concept with which our modern society is only just coming to terms.)

Yet, conservation ethics were already forming in those early days. Small national parks were gazetted in the first years of this century — Witches Falls (Tamborine), the first national park in Queensland, was declared in 1906 and Cunningham's Gap followed in 1909. Lamington was declared in 1915 and the reserves of the Scenic Rim (those mountains encircling the region from Blackall through Main Range, Mt Barney and Lamington to Springbrook) began to take shape.

(This was impressive because Australia as a nation was then only a few years old and only about 35 years behind the United States in recognising that conservation was of national importance. The world's first national park, Yellowstone, had been declared in 1872.)

The foresight of the early conservationists lives on in their legacy — the parks and reserves of Queensland — but this may not be enough to ensure the future. In recent years, valued bushlands have continually been replaced by roads, homes, agriculture and industry. The remaining natural areas are fragmented and their borders are not always strong enough to hold back the encroachment of modern society.

Everywhere in the Greater Brisbane Region, the demand for land is increasing. The developed areas of Brisbane, Logan, the Gold Coast, Pine Rivers, Ipswich and Redland are extending in all directions. It is not inconceivable that within a few decades a giant conurbation will stretch from Noosa on the Sunshine Coast to Murwillumbah in northern New South Wales, south-west to Beaudesert and west to Laidley in the Lockyer Valley.

It has already happened in America and in Europe where connected urban sprawl swallows towns and cities one after the other. Night-time images from the space shuttle show there are

Rainforest logging, 1946 JOL

Stradbroke Island, Christmas 1885 JOL

fewer and fewer areas around the world without street lighting. Humans damage and destroy in a short space of time that which has taken eons to evolve — we need to take stock of what we have before it is too late.

Until recently bushland was being lost at an alarming rate in South-east Queensland. In the 15-year period to 1989, one-third of the remaining bushland was cleared.[5] While the rates of clearance in Brisbane have slowed, this is not true for the rest of South-east Queensland. If high clearance rates continue, then in a very short space of time — perhaps only 10 to 20 years — there will be no freehold "bush" left in some areas.

What little remains will be sheltered inside the reserve system and even this protected isolation may not guarantee the survival of many plant and animal species. Many reserved areas can become the "path of least resistance" for infrastructure such as major arterial roads.

Additionally, some animals need large areas of often specialised habitat to survive. Isolation and urban edge effects — pollution, predation by domestic animals and road kills caused by traffic — are the beginning of the slow march towards extinction for many. Similarly, sensitive plant species need buffers against the often detrimental effects of development. These effects need not be direct. In some cases they are secondary or cumulative, but the results are no less severe. Natural ecosystems are complex and interlinked — they cannot be considered in isolation.

To retain and maximise the region's biodiversity, large, intact habitats are needed and these habitats should be linked to each other. Where this is not possible, the connections need to be re-established over time. Imagine a jigsaw — it is better to have an almost complete puzzle with only a few pieces missing than to have just one or two separate pieces on a bare board.

Yellow Carabeen, Tamborine, 1931 JOL

As the year 2000 approaches, most people realise that some bushlands have to remain — not only for the plants and animals and environmental quality, but for our own peace of mind, enjoyment and relaxation. Natural areas are "breathing spaces" important for our well-being, as recreational escapes from the strains of city life and as vicarious emotional supports (we feel good knowing they exist). They improve the quality of our lives and it is only now, as we approach the boundaries of the last "wild places", that we realise how important they are.

Greater Brisbane has been fortunate —it is not too late to preserve our natural heritage for the generations to come — all we have to do is want it hard enough. Ironically, as modern life becomes more separated from the natural world, more people are being drawn to it. Perhaps one way to secure the future of our "wild places" is to make them financially attractive. Ecotourism — where tourists pay to be taken to outstanding natural areas and learn about the geography, plants, and animals — may be an option.

Ecotourism is worth many billions of dollars globally each year and it is the fastest growing component of the tourism industry. Sustainable ecotourism provides a monetary incentive to protect native plants and animals (thereby allowing us to continue to enjoy them); it can boost local economies and provide employment; and perhaps most importantly, it has the potential to maintain life-support systems for the whole planet.

But before we can plan, we need to know what's beyond the backyard — let's go and explore . . .

Stephen Poole

HABITATS

The natural habitats of the Greater Brisbane Region are as varied as the plants and fauna they support. Although much has been lost, the region still has rainforests, eucalypt and paperbark forests and woodlands, heathlands, dune communities, and freshwater and tidal wetlands.

Brisbane Forest Park

RAINFORESTS

When Europeans first ventured up the Brisbane River in the 1820s they saw before them imposing rain-forest-clad banks with dense growth of ferns, vines and towering trees. In the lowlands beyond the river, rain-forests were more prevalent than they are now and stretched across much of the region. On hills and mountains, they formed vast scrubs.

Today the remnant rainforests of South-east Queensland and northern New South Wales are rated second (on a national basis) in importance only to the Wet Tropics of Far North Queens-land. This is not just for their diversity of plants and animals. Like all rain-forests, their plant species have links with the ancient vegetation of Gond-wana, the super continent which existed around 135 million years ago.

Most of the surviving rainforests around the Greater Brisbane Region are in higher areas — particularly mountains and plateaus — on deep, rich basaltic soils. Lowland rainforests are now confined to the valleys of the McPherson Ranges with small patches on creeklines and river floodplains

such as at Bahr's Scrub near Beenleigh, Bell's Scrub at Lawnton and Ward's Scrub at Samford.

In rainforests, tall trees grow close together and the tops of the trees form a dense, closed canopy that filters out much of the sun. (Look up into the canopy next time you visit a rainforest and see how little light penetrates. This is one of the reasons rainforests are cooler in summer.) At mid-levels and on the ground, there are palms, vines and lianes, ferns, lichens and mosses. Epiphytes, plants that live in beneficial associations with other plants, are common.

Opposite: Brisbane Forest Park **Above:** Main Range, Teviot section

An interesting quirk of nature is the association between the **Giant Stinging Tree** (*Dendronicide excelsa*) and the **Cunjevoi** (*Alocasia brisbanensis*), shown left. Looking up into the rainforest canopy, the Stinging Tree can be identified by its large, heart-shaped leaves which are often skeletonised by insects. The leaves of this tree should never be touched because they are covered in fine hairs which inject a substance that can cause severe and prolonged pain. The Cunjevoi, which is also known as Elephant's Ear, almost always grows near the Stinging Tree but at ground level. It is a highly toxic plant, but its sap and leaves were ground to a paste by the Aborigines and used as an antidote to painful encounters with the Stinging Tree. (This is similar to the Stinging Nettle in Europe and the nearby antidote, the Burdock leaf.)

Rainforests have a distinctive fauna. Most of the insects, birds, reptiles, mammals and other animals found there do not live anywhere else. The Elf Skink (*Eroticoscincus graciloides*) for example, lives mainly in the lowland rainforests of southern Queensland, where it is rare and difficult to find. Such habitat dependence demonstrates the importance of rainforest as an ecosystem.

In the Greater Brisbane Region there are four rainforest types:

SUB-TROPICAL RAINFOREST

This is the most common rainforest and has the greatest species diversity. It is associated with the region's subtropical climate and grows on fertile soils on mountains and plateaus such as at Mt Glorious, Tamborine, Lamington and on the ranges around Maleny, as well as in a few isolated lowland areas. (Rainfall is generally 1500 mm or above per annum.)

Sub-tropical rainforest is characterised by massive trees which often have prominent buttressed roots and large leaves. Species include Strangler Fig (*Ficus watkinsiana*) with its unusual "latticed" growth pattern (see p. 12), Moreton Bay Fig (*F. macrophylla*), Yellow Carabeen (*Sloanea woollsii*), Wheel of Fire Tree (*Stenocarpus sinuatus*) and the Black Bean or Moreton Bay Chestnut (*Castanospermum australe*). The Black Bean is easily identified by its seed pods which are the largest in the forest — up to 15 cm long. During the early years of European settlement, Red Cedar (*Toona ciliata*) was prized by timbergetters and consequently few large specimens remain, even in relatively secure forests.

High in the canopy, epiphytic Elkhorn (*Platycerium bifurcatum),* Staghorn (*P. superbum*) and Bird's Nest (*Asplenium australasicum*) ferns grow on the upper trunks and branches of the larger trees, along with many orchids.

In the mid-levels, palm groves, particularly of Bangalow or Piccabeen Palm (*Archontophoenix cunninghamiana*) and scattered stands of Cabbage Tree Palm (*Livistona australis*) are common, as are tree ferns.

Various wiry vines and larger, woody lianes twine their way round trunks and branches, and ferns and a variety of other plants grow at ground level. Shield Ferns (*Lastreopsis* spp.), Native Raspberries (*Rubus* spp.), and Christmas Orchid (*Calanthe triplicata*) are typical understorey plants.

What is considered to be one of the largest mosses in the world, *Dawsonia superba,* grows in the sub-tropical rainforests of the Greater Brisbane Region. This plant, which looks like tiny pine seedlings, often grows to 10 cm long.

Above and below: Tamborine

The Strangler Fig has one of the most interesting growth patterns of any plant. Seeds lodge in the fork of trees or other suitable places and when they germinate, send aerial roots down to the ground. These roots then repeatedly branch, meshing together and forming a lattice structure around and up the host tree. The fig then slowly "strangles" the host tree over many years by competing with it for nutrients and for light. The host tree gradually dies leaving the hollow "framework" of the fig. Sometimes fig roots can be seen twining round rocks and boulders and extending for enormous distances — up to 100 m.

DRY RAINFOREST

Dry rainforest tends to grow on fertile and moderately fertile soils from ridges to mountain tops and in the lowlands, from ridges to drainage lines in areas of lower rainfall (up to 1200 mm per annum).

This forest type is also known as vine scrub. The forests experience a relatively dry season during late winter and spring. Good examples of dry rainforest exist at Mt Mee, Lamington and Main Range.

The species in dry rainforests are as rich as, or even more so, than those found in moister forests. The canopy trees are still tall, though generally smaller than those in sub-tropical rainforest and include "emergent" conifers (*Araucaria* spp.) that protrude above the canopy.

Dry rainforest trees generally have smaller leaves and may lack the distinctive buttressing of sub-tropical species. There are fewer epiphytic plants, palms and ferns and the mid-layer vegetation often has numerous vines and prickly shrubs.

Hoop Pine (*Araucaria cunninghamii*) is a characteristic emergent tree of dry rainforest in this region, as is Bunya Pine (*A. bidwilli*) which is more restricted in its distribution and at its southern limit around Mt Mee. Hoop Pine was a major economic base of the early European settlement in the Greater Brisbane region and was called Colonial Pine (see p. 110). Other dry rainforest species include Crow's Ash (*Flindersia australis*), Crow's Apple (*Owenia venosa*), and Red Kamala (*Mallotus philippensis*).

Around Brisbane the regrowth of dry rainforest is indicated by the presence of pioneer species such as acacias, native crotons and Red Ash (*Alphitonia excelsa*) or Soap Tree as it is sometimes called.

Opposite: Jimna State Forest **Inset:** Left, Seed pods of Crow's Ash; **Right**, Mt Mee

Opposite and above: Warm temperate rainforest, Lamington National Park
Bottom right: Cool temperate rainforest

WARM TEMPERATE RAINFOREST

This forest generally occurs at higher altitudes than sub-tropical rainforest and is restricted to poorer soils at Lamington and Springbrook. As few as three species of trees can form the canopy layer and generally the canopy is not as high as sub-tropical rainforest. Rainfall is comparable to that of sub-tropical rainforest.

The tree trunks are narrower, of more uniform size than sub-tropical species and are often covered with whitish lichen. Coachwood (*Ceratopetalum apetalum*) typically dominates this rainforest type which also has large numbers of tree ferns. Many of the understorey shrubs have prickly leaves.

COOL TEMPERATE RAINFOREST

Only a few small patches of cool temperate rainforest grow in the Greater Brisbane Region and these are in the high rainfall areas of the McPherson Ranges. Cool temperate rainforest occurs at the highest altitudes on exposed crests and also on some of the higher, colder valleys. This simple form of rainforest is dominated by Antarctic Beech (*Nothofagus moorei*) and tree ferns. The upper branches of the beeches are often festooned with mosses and lichens which gain much of their moisture from their frequent exposure to low cloud (see p. 159).

EUCALYPT FORESTS AND WOODLANDS

E ucalypts, our most familiar trees, are synonymous with the Australian "bush". Eucalypt communities are rich in plants and animals. The forests within the Greater Brisbane Region, for example, support more than 30 eucalypt species and even very small areas only a few hectares in size may have as many as eight different species.

Forest Red Gums (*Eucalyptus tereticornis*), ironbarks (*E. siderophloia, E. crebra* and *E. fibrosa*) and mahoganies (*E. carnea, E. acmenoides*) grow at most sites, but the composition varies endlessly with soil type, location, slope, moisture, frequency of fire and other factors in the complex web of ecological interactions that mould our forests. However, in drier areas and on poorer soils, the open forest structure gives way to woodland where the trees are further apart and tend to be shorter.

Species allied to the eucalypts include Brush Box (*Lophostemon confertus*), Bloodwoods (*Corymbia* spp.) and Apples (*Angophora* spp.) and can be found from dry ridgelines to wet eucalypt forests throughout the region.

Eucalypt communities are home to some of Australia's better known animals — koalas, kangaroos and wallabies. The leaves and blossoms of eucalypts are food for bats, possums, insects and birds. As in most other habitats, the mammals in eucalypt forests are mostly nocturnal and hard to find without a spotlight. Reptiles are also abundant ranging from small skinks and geckoes to large goannas and pythons.

Opposite: Mt Mee State Forest **Above:** Brisbane Forest Park

The two types of eucalypt forest around Brisbane are:

DRY EUCALYPT

Dry eucalypt (or dry sclerophyll) forest is the typical "bush" of the Greater Brisbane Region. Most eucalypt species can be found in dry forest with different combinations occurring on different soils and at different altitudes.

For example, in sub-coastal areas with poorer soils, the distinctive Spotted Gum (*Corymbia variegata*) grows on ridgelines. Spotted Gum can be recognised by its very long, broad leaves and indented, but smooth pinkish bark. On good soils with more sheltered, moister aspects, the Grey Gum (*E. propinqua*) and Tallowwood (*E. microcorys*) are dominant. Tallowwood has dark green leaves and stringy bark which is orange or red-brown. Forest Red Gum grows along creek lines on better alluvial soils.

The understorey in dry eucalypt forests can be either shrubby or grassy. Shrubby understoreys contain a variety of plants including banksias, wattles, pultanea, hakeas, daviesia and hoveas. Better known species are Coast Banksia (*Banksia integrifolia*), Brisbane Wattle (*Acacia fimbriata*) and the grass tree, *Xanthorrhoea johnsonii*.

Most dry eucalypt forests are subjected to periodic bushfires and fire was a land management practice of the Aborigines for many thousands of years. Where fires are frequent, the understorey is usually grassy and dominated by Kangaroo Grass (*Themeda triandra*), a tall soft grass with oaten heads, or Blady Grass (*Imperata cylindrica*).

A heath understorey can develop where the soil is water-logged and low in nutrients such as on sandstone soils. Typical plants are banksias, tea-trees (*Leptospermum* spp.), paperbarks (*Melaleuca* spp.), small grass

Opposite: Main Range, Spicer's Gap

In relatively intact eucalypt forests, there are large numbers of older trees which have hollows in the branches and trunks. These are "habitat" trees and they are vital to the long-term survival of many animals. As the name suggests, the hollows provide shelter and nesting places for birds, bats and arboreal mammals such as possums and gliders. Scratch marks on the bark and droppings (scats) around the base of the tree are evidence that the tree is being used. It takes many years for trees to develop hollows. Where forests have been heavily logged and hollows are lacking, so too are the animals that depend on them. It is not uncommon to see new residential developments which have been selectively cleared with only a few, young straight-growing gums left. While this appears to be environmentally friendly planning, these trees are of no use to nesting animals. As well, nesting trees tend to be removed when they occur near walking tracks in forests because of the danger of falling limbs. The solution is not to cut down the tree but to move the walking trail. — SP

trees without trunks and low stunted eucalypts. Wildflowers — boronias, peaflowers and the like — are prevalent in heath understoreys

As with animals, plant species face similar threats from habitat destruction. Two significant species of dry eucalypt woodlands are Planchon's Stringybark (*E. planchoniana*) and Bailey's Stringybark (*E. baileyana*). These trees exist in only a few locations around the Greater Brisbane Region on sandstone soils such as near Plunkett, Toohey Forest, and the Helidon Hills.

Smudgee (*Angophora woodsiana*), Smooth-barked Apple (*A. leiocarpa*) and *A. subvelutina* are also found in woodlands. These trees are distinguished from eucalypts by their fragile seed capsules which have four equally spaced ridges.

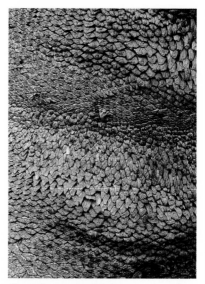

Above: Grass tree trunk
Below: Dry eucalypt forest, Mt Barney

Above: Wet eucalypt forest, Lamington National Park

WET EUCALYPT

Wet eucalypt (wet sclerophyll) forest grows on fertile soils with relatively high rainfall (1500 mm per annum) such as in protected gullies and at the margins of rainforest, as at Mt Nebo.

Wet eucalypt forest is characterised by sometimes very tall trees such as the Flooded Gum (*E. grandis*), Turpentine Tree (*Syncarpia glomulifera, S. verecunda*), Sydney Blue Gum (*E. saligna*), White Mahogany (*E. acmenoides*) and Brush Box.

The understorey of wet eucalypt forest includes many rainforest plants, particularly palms and tree ferns and the distinction between wet eucalypt and rainforest can become blurred. Some of the best places to see well-developed wet eucalypt forests are on the Tamborine Plateau/Tamborine Village Road as it passes through the Joalah section of Tamborine National Park and also around Mt Glorious.

Sydney Blue Gum bark *Eucalyptus saligna*

HEATHLANDS

Heathland usually refers to low, shrubby vegetation which grows on infertile soils. For example, on the islands of Moreton Bay heaths grow on sand. Elsewhere in the region, heaths can be found in water-logged areas, often on sandstone soils. Heaths can be wet or dry, but here they are considered as a single type. Heathlands that occur near the coast are often referred to as wallum heath.

Although plant diversity can be high, heathland shrubs often have stunted growth and small, tough leaves. Eucalypts in heathland usually have a "mallee" form. The word "mallee" comes from the Mallee district of southern Australia and the trees are usually only a few metres high and have multi-stemmed trunks.

Common heathland species in the Greater Brisbane Region include grass trees (*Xanthorrhoea* spp.), banksias, callistemons, small tea-trees (*Lepto-*

spermum spp.), boronias, persoonias and members of the pea family (*Pultenaea* and *Dillwyinia* spp.).

Other small areas of heathland known as montane heath also occur in the Greater Brisbane Region. These are restricted to the tops of mountains such as the Glasshouses, Mt Barney and Mt Ballow and on the acid soils at Springbrook and Lamington (Dave's Creek). Like lowland heaths, the plants are shrubby and have small, hard leaves.

Due to their isolation, these montane heaths often have rare or restricted species found only on these localities.

In late winter and spring, heathlands explode into bloom and hordes of honeyeaters, lorikeets and bats are lured to the nectar-laden flowers. One of the most striking birds is the black, white and yellow, White-cheeked Honeyeater (*Phylidonyris nigra*). Other heathland inhabitants include bandicoots, quail, rodents and lizards.

Opposite: Wallum heath, Moreton Island
Below: Montane heath, Lamington National Park

DUNE COMMUNITIES

A long beaches, the plant life is very distinctive. On foredunes — those closest to the beachfront just above the high tide mark — the sand is very visible and the vegetation sparse. Nevertheless the plants that grow here — strand vegetation — help stabilise the dunes and prevent them being blown or washed away.

Creepers such as Beach Morning Glory (*Ipomoea pes-caprae* subsp. *brasiliensis*), and grasses such as Beach Spinifex (*Spinifex sericeus*) are common plants of the foredunes. Sometimes the dune vegetation is backed by low forests of Beach She-oak (*Casuarina equisetifolia* var. *incana*) and older dunes further inshore, such as on Bribie, Moreton and Stradbroke Islands, even support eucalypt forests.

Heath species are found on the more exposed slopes of higher, coastal dunes while Pandanus (*Pandanus tectorius*) with its enormous dangling leaves is a distinctive plant of rocky headlands such as Point Lookout on Stradbroke Island and Burleigh Head. It is around rocky outcrops like these that the sand islands of Moreton Bay originally formed.

Opposite and above: Moreton Island

WETLANDS

A wetland is essentially an area of land that becomes either permanently or temporarily inundated by water and typically is low-lying and swampy. There are two types of wetland — freshwater fed by creeks, rivers and rainfall; and tidal, which as the name suggests, are under seawater for at least part of the day.

Paperbark forests and woodlands (which are also referred to as paperbark swamps), may only be inundated for relatively short periods on a seasonal basis. However there is usually a high water table which keeps the sub-soil water-logged and therefore unsuitable for many other plant species.

Wetlands are important ecosystems. They are known to act as biological filters for pollution and can reduce contamination of the ocean through the breakdown of pollutants such as sewage.

Although they are not included in the habitat listings of this book, man-made water impoundments such as farm dams and water supply reservoirs can form important habitats for bird life and provide water for native mammals during times of drought. The Ewen Maddock Dam near Landsborough is well known for its bird life and a hide has been constructed for non-intrusive observation.

Many wetlands in coastal regions have acidic soils and by coincidence these are usually areas where development is occurring. When the wetlands are cleared, development infrastructure such as bridges and concrete foundations are built. The mechanical action of the development activates sulphuric acid in the soil which gradually eats into these structures. As well as the detrimental effect on building foundations, the sulphuric acid can be washed into waterways where it can damage mangrove and intertidal areas and subsequently effect animal populations including fish stocks.

Opposite: Eighteen Mile Swamp, North Stradbroke **Above:** Blue Lake, North Stradbroke

FRESHWATER

SWAMPS

Freshwater swamps are inundated for much of the year. Rushes and sedges dominate the vegetation particularly at the edges of swamps where they form dense stands. Twig Rush (*Baumea articulata*), Spikerush (*Eleocharis equisetiana*) and sedges such as *Carex appressa* are common throughout the region.

Swamps may also support several species of waterlillies and other floating plants. Eucalypts, paperbark trees, and ferns such as Coral Fern (*Gleichenia* spp.) and Bungwall Fern (*Blechnum indicum*) may border swamps and in some cases form the understorey.

Some swamps in the Greater Brisbane Region have a base of peat formed over thousands of years by decaying vegetation and animal matter. This makes the water brackish and reduces the oxygen content of the water.

Flinders Swamp on Stradbroke Island and Bulwer Swamp on Moreton Island are two of the largest freshwater swamps in the region and there are also small swamps on Bribie Island and a few on the mainland.

Bribie Island

LAGOONS AND LAKES

Lagoon systems (which include billabongs) fill after rain and only occasionally dry out. They often form along watercourses through dry eucalypt forest. Unfortunately, the Greater Brisbane Region has lost most of its original lagoon systems due to land clearance and filling. Among those that survive are Sandgate Lagoon, the Illaweena Lagoons at Karawatha and the lagoons of Upper Buhot Creek.

The edges of lagoons often have dense covers of ferns, rushes and sedges interspersed with paperbark trees. Eucalypt forest may also grow right to the water's edge.

The best examples of freshwater lakes in the Greater Brisbane Region are on Moreton and Stradbroke Islands. Again, these lakes are fringed by sedges and rushes including the Grey Sedge (*Lepironia articulata*) and Twig Rush and sometimes by paperbark trees and Bungwall Fern.

Above: Buckley's Hole, Bribie Island
Below: Tortoise Lagoon, North Stradbroke Island

Lagoons and lakes are a haven for wildlife such as tortoise, freshwater snakes, platypus and mammals like the Swamp Rat (*Rattus lutreolus*) along with crayfish (yabbies), freshwater sponges, insects and of course, frogs

Mention should be made of the temporary ponds and lagoons that form after rain in drier habitats. Although they soon evaporate, they are vitally important for frogs to breed.

PERCHED LAKES AND SWAMPS

On Moreton and North Stradbroke Islands, lakes have formed in depressions among the higher sand dunes. These "perched" lakes are substantially above the watertable and water has difficulty draining away because of an almost impervious layer of accumulated decaying organic matter and other debris. (Traditional lakes are also formed in depressions but in this case, the watertable is very close to the surface.)

Blaksley and Black Snake Lagoons on North Stradbroke Island are examples of a perched swamps.

The decaying organic matter makes the water in these lakes and swamps acidic and a number of animals have adapted to these conditions — fish, shrimps and prawns, frogs and insects. Some species are confined to these limited habitats and as a result many are considered rare or vulnerable, including the Wallum Sedgefrog (*Litoria olongburensis*), Cooloola Sedgefrog (*L. cooloolensis*) and the Oxleyan Pygmy Perch (*Nannoperca oxleyana*).

Brown Lake, North Stradbroke Island

North Stradbroke Island

PAPERBARK FORESTS AND WOODLANDS

Along the eastern fringes of the Greater Brisbane Region, much of the land is flat and composed of alluvial sediment. After heavy rains, this soil becomes saturated and swamps form. Most eucalypts cannot tolerate this water-logged environment but Paperbark Tea-trees (*Melaleuca quinquenervia*) thrive in it.

Paperbark forests and woodlands were important to the Aborigines, yielding supplies of Bungwall Fern, of which the starchy underground stems were a staple food. Bark stripped from the tea-trees provided roofing for huts.

European settlers, however, scorned the mosquito-ridden "swamps" and felled vast tracts for pasture and agriculture. More recently, the dwindling paperbark stands have come under threat of development for housing. In South-east Queensland more than half the paperbark forests were cleared between 1974 and 1989.

Although scattered paperbarks often grow along the edges of waterways, only small areas of forest and woodland now remain at the Deagon Wetlands, Serpentine and Native Dog Creeks (which overlap the boundaries of Logan City and Redland Shire), and Boondall and Tinchi Tamba on the north of Brisbane.

Paperbarks are easily identified by their very finely layered bark which peels off untidily. During late summer and autumn, the trees have masses of yellow flowers.

Depending on the degree of water-logging, few other plant species may grow in paperbark communities. The paperbark itself can grow as a tall or short tree, depending on soil conditions. Blady Grass (*Imperata cylind-*

rica) is common in the understorey of drier forests, particularly where they have been burnt, while Bungwall Fern, Swamp Rice Grass (*Leersia hexandra*) and various sedges grow along channels. Silkpod (*Parsonia straminea*) is a common vine. On the offshore islands, paperbarks can grow as part of a wallum community and the understorey is more diverse.

Two eucalypts sometimes found growing among the paperbarks are the Swamp Mahogany (*Eucalyptus robusta*) and to a lesser extent, the Forest Red Gum (*E. tereticornis*). A related species which often grows with Forest Red Gum is the Swamp Box (*Lophostemon suaveolens*).

Melaleuca quinquenervia is not the only paperbark to form a distinctive habitat. Around Boonah and southwest of Ipswich, there are very small remnants of Swamp Tea-tree forest (*M. tamariscina irbyana*), a once extensive habitat type that is now rare.

The surviving paperbarks forests are important refuges for animals. Paperbark flowers secrete copious nectar for many birds and bats, especially lorikeets, honeyeaters and flying foxes. A special visitor to the blooms is the rarely-seen Queensland Blossom Bat (*Synconycteris australis*), only 5 cm long.

The swamps that form after rain attract hordes of breeding frogs which in turn attract snakes such as the harmless Keelback (*Tropidonophis mairii*) and Carpet Python (*Morelia spilota variegata*). Other denizens of paperbark forests are gliders, kangaroos and echidnas.

Butterflies are often abundant among paperbarks because of the prolific flowering. The Black and White Tiger (*Danaus affinis*) and rare Brown Soldier (*Junonia hedonia*) are largely restricted to this habitat as their food plants only grow in coastal swamps.

Opposite: Paperbark swamp, Daisy Hill

TIDAL

MUDFLATS, MANGROVES AND SALTMARSHES

Along shorelines, particularly in Moreton Bay and along Pumicestone Passage, mangroves form dense stands up to the high water mark. At low tide, mudflats are exposed.

These bare expanses of mud and pools look barren at times, but in summer they are often rich in bird life. Wading birds can be seen using their long beaks to probe deep into the mud in search of molluscs, crustaceans and other invertebrate animals such as worms.

Some mudflats support lush pastures of seagrass, a staple food of the dugong herds which are at their most southerly limit in Moreton Bay. The plants are sometimes exposed by very low tides.

Mangroves are trees which can live in saline conditions and which are partially submerged at high tide. Mangroves can form large continuous stands or fringing forests along tidal creeks and rivers.

In the past mangroves were reviled as mosquito breeding grounds and targeted for reclamation to be buried under sports fields and canal estates.

Deception Bay

Today the importance of mangroves and mudflats as nurseries for prawns, crabs and particularly fish is widely recognised. Although it is known that the removal of mangroves deplete the fish stocks, it is still uncertain how severely this will have affected Moreton Bay.

At low tide, mangrove forests are characterised by many aerial roots. These aerial roots not only stabilise and support the tree in the dense, permanently wet silt in which mangroves grow, but they also help the plant obtain oxygen.

Stilt roots

Aerial roots which arch downwards from the trunk are called stilt or prop roots. Pencil-like roots which stick up out of mud are called pneumatophores and are characteristic of the Grey Mangrove (*Avicennia marina*), the most abundant species of the Greater Brisbane Region. Last century, convicts burned the wood of the Grey Mangrove to obtain soda for soap making.

Other species are the River Mangrove (*Aegiceras corniculatum*), Milky Mangrove (*Excoecaria agallocha*), Spurred Mangrove (*Ceriops tagal*), Orange Mangrove (*Bruguiera gymnorhiza*), Red Mangrove (*Rhizophora stylosa*) and Black Mangrove (*Lumnitzera racemosa*). The Black Mangrove occurs only as far south as Moreton Bay while the Spurred Mangrove reaches its southern limit at Tallebudgera.

Many animals aside from fish rely on mangroves. Stands of Grey Mangrove within the Greater Brisbane Region support several large roosts of Grey-headed and Black Flying Foxes (*Pteropus poliocephalus, P. alecto*).

Honeyeaters are frequent mangrove visitors as are rodents such as the rare Water Mouse (*Xeromys myoides*) and various invertebrates (see p. 117).

Saltmarshes are usually found just inland from the mangroves. They are characterised by low shrubs including Samphire (*Halosarcia indica, H. pergranulata*), Seablite (*Suaeda australis*) and the Berry Salt Bush (*Enchylaena tomentosa*). Amongst these shrubs grow grasses and sedges such as Saltwater Couch (*Sporobolus virginicus*) and the Maritime Rush (*Juncus maritima*).

Saltmarshes vary in extent with some areas being inundated at each high tide and others remaining above water until very high or spring tides. Naturally, saltmarsh vegetation is extremely salt resistant and at low tide bare patches between the vegetation look not unlike clay/salt pans. Saltmarshes are less diverse in the tropics and sub-tropics than they are further south.

Above: Saltmarshes, Tinchi Tamba Wetlands
Below: Boondall Wetlands

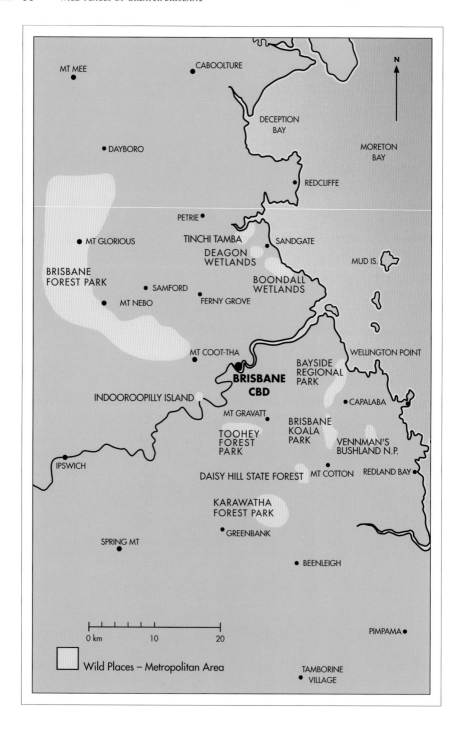

MT MEE

CABOOLTURE

N

DECEPTION
BAY

DAYBORO

MORETON
BAY

REDCLIFFE

PETRIE

MT GLORIOUS

TINCHI TAMBA

SANDGATE

DEAGON
WETLANDS

MUD IS.

BRISBANE
FOREST PARK

SAMFORD

BOONDALL
WETLANDS

MT NEBO

FERNY GROVE

MT COOT-THA

WELLINGTON POINT

**BRISBANE
CBD**

BAYSIDE
REGIONAL
PARK

INDOOROOPILLY ISLAND

CAPALABA

MT GRAVATT

BRISBANE
KOALA
PARK

TOOHEY
FOREST
PARK

VENNMAN'S
BUSHLAND N.P.

IPSWICH

DAISY HILL STATE FOREST

MT COTTON

REDLAND BAY

KARAWATHA
FOREST PARK

SPRING MT

GREENBANK

BEENLEIGH

0 km   10   20

PIMPAMA

Wild Places – Metropolitan Area

TAMBORINE
VILLAGE

WILD PLACES — METROPOLITAN AREA

Toohey Forest

BRISBANE FOREST PARK

THE BUSH AT BRISBANE'S DOORSTEP

Imagine exploring a pristine rainforest of towering buttressed trees, discovering a clear, deep rock pool with cascading waterfalls, or being able to walk almost endlessly through a sea of green forests and camping by a remote mountain stream.

All this and much more is possible within a short drive of Brisbane. Many people would be surprised to learn just how much forest and wildlife diversity is there, right on their doorstep. Brisbane Forest Park can provide many opportunities for anyone with an interest in nature and a love of the great outdoors.

Brisbane Forest Park is one of the biggest and most accessible reserves of its type so close to a major city anywhere in the world. Beginning at Mt Coot-tha, the forest covers 28,500 ha and stretches continuously for 70 km along the Taylor and D'Aguilar Ranges extending out to Wivenhoe Dam in the west. The park provides a scenic, green backdrop to the city and a nature-based bush escape for its residents. Already some 2 million people visit the

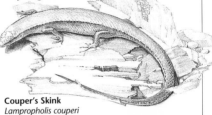

LOCALITY GUIDE

Location: Park Headquarters 15 km west of Brisbane GPO; 20 mins' driving.

Access: Via Mt Nebo and Mt Glorious Roads from the east and Northbrook Parkway from the West.

Facilities: Numerous picnic areas; lookouts; walking tracks; headquarters complex at The Gap includes, visitor centre and freshwater wildlife display.

Restrictions: No domestic pets; no off-road driving/motorcycling; permits required for camping, horse-riding and cycling.

Couper's Skink
Lampropholis couperi

park each year and yet with sensitive management and a responsible attitude by visitors Brisbane Forest Park can continue to serve as a valuable community asset.

Opposite: Love Creek **Above:** Early morning mist

Above: Northbrook area **Opposite:** Near Wivenhoe Outlook

THE MAKING OF A PARK

The present landscape of the park is a legacy of a long and sometimes violent geological history. The result is a landscape of prominent, steep-sided valleys and gorges, occasional isolated peaks and a long, undulating north-south plateau.

The comparative ruggedness of the area has been largely responsible for preserving the D'Aguilar Ranges in their present, relatively natural state. During the early years of European settlement much of this region was overlooked for farming, being considered too difficult to develop. Brisbane's first water storage, Enoggera Reservoir, was constructed in these foothills (at The Gap) in the 1860s, and as a consequence much of the surrounding catchment gained some form of protection.

Over the ensuing years construction of two more reservoirs — Gold Creek and Lake Manchester — and the allocation of land to State Forest and National Park reserves, effectively locked out any other major development. However during the past 100 years there has still been considerable activity around the D'Aguilar Ranges.

The area served as a valuable timber resource for the Brisbane region, and even today small quantities of native hardwoods are removed on a sustainable logging basis. During the Depression, a few hardy souls attempted to eke out an existence from gold prospecting, but with very limited success. Small scale farming in the more fertile areas around Mt Nebo and Mt Glorious led to the establishment of these two villages which today serve as ideal retreats removed from, but still in sight of, the hurly-burly of the city.

Greater environmental awareness during the 1970s gave rise to growing community interest in providing for better management of the natural, aesthetic and recreational values of the D'Aguilar Ranges. In response to this community pressure the State

Wet eucalypt forest

Government established the Brisbane Forest Park Administration Authority in 1977, thus creating Queensland's first coordinated conservation area.

The main role of the authority is to provide for the recreational use of the park while at the same time recognising the other important roles of its National Parks, State Forests and Brisbane City Council Reserves. Brisbane Forest Park is a truly multi-purpose reserve, not only providing for nature-based recreation, education and tourism but also water catchment and storage, timber and honey production, and nature conservation.

There are many intrinsic values to be gained from a park of this size so close to a large population centre. Those we take for granted include fresh air and

Giant Panda Snail
Hedleyella falconeri

unspoiled scenery, an effective buffer and backdrop to the urban sprawl of the city and a continuous wildlife corridor around virtually all of the western and northern suburbs. Many of the wildlife species that inhabit our suburban gardens, and those rarer species that occasionally drop by, only exist due to the proximity of nearby bushland which provides a necessary safe haven and additional food sources.

THE PARK FLORA

Woodlands and dry eucalypt forests predominate on the drier, shallower soils of the Park foothills. Spotted Gum (*Corymbia variegata*) and Narrow-leaved Ironbark (*E. crebra*) are two of the main species. Fire is a frequent influence on these forests and during dry spring weather controlled burns may be undertaken. Small areas of heathlands and ridges dominated by grass trees add variety to the landscape.

Along valleys such as Enoggera Creek, Northbrook Creek and the South Pine River, remnant pockets of lowland rainforest occur. Once much more widespread in the Greater Brisbane

Magnificent Spider
Ordgarius magnificus

Region, these forests were logged and cleared providing timber such as Hoop Pine (*Araucaria cunninghamii*) for many of the early Queenslander houses. These few remaining pockets protect rare species such as the Broad-leaved Whitewood (*Atalaya multiflora*), Brown Myrtle (*Choricarpia leptopetala*) and the spectacularly buttressed Giant Ironwood (*C. subargentea*).

At higher altitudes the forests become more complex due to the increased rainfall and generally deeper soils. Species such as Grey Gum (*Eucalyptus propinqua*), Pink Bloodwood (*Corymbia intermedia*) and their close relative the Brush Box (*Lophostemon confertus*), tend to dominate these mid-

White Cedar *Melia azederach*

A PARK FOR THE PEOPLE

Brisbane Forest Park provides many opportunities and a well developed range of facilities that enables everyone to get out and enjoy the Australian bush. There are 16 separate recreation areas across the park providing picnic facilities, spectacular lookouts, self-guided walking tracks and a variety of opportunities for bushwalking and camping.

These facilities have been developed without encroaching on the main body of the park, by adopting a policy of concentrating human impact at appropriately spaced recreation nodes along the public road system that skirts the park. As a result, most of the park is still largely intact and only accessible for those prepared to be more adventurous and leave their car behind. The park boasts a network of forestry roads which, while closed to private vehicular access, provide the perfect venue for safe and relaxing horse-riding and all terrain cycling.

Visitors to Brisbane Forest Park are not just left to their own devices. Through its Education Program school children, with specialist Ranger input, can use the park as an outdoor living classroom. The "Go Bush" program is an imaginative way for individuals, community groups and businesses to explore the world of nature and forest recreation all year round under the expert guidance of park rangers and volunteers

The programs run by Brisbane Forest Park provide an ideal way for anyone with an interest in nature to tap into the expertise and facilities on offer. It also allows the Authority to better manage the park by having more influence on visitor usage, thereby reducing the risk of degradation and developing a sense of community care, responsibility and pride for one of the greatest natural assets in the Greater Brisbane Region.

altitude forests, but occasional rainforest species are evident and a thick understorey of ferns, vines and shrubs is present.

Fire still plays an important role in these forests, and while less frequent it is more intense and the occasional destructive wildfire influences the boundaries between eucalypt forest and rainforest. Evidence of the shifting balance can be seen on walks around Mt Nebo where remnant eucalypts can be found in the centre of rainforests. At the rainforest edge, pioneer species such as Bleeding Heart (*Omalanthus nutans*) and a proliferation of vines try vigorously to gain a foothold until the next fire decimates them and allows the eucalypts to re-invade, in a cycle played out several times each century.

On the highest parts of the range where the rainfall is two-thirds greater than that in the foothills, moist sub-tropical rainforest grows, particularly on the

rich, basaltic soils north of Mt Glorious. Huge Strangler Figs (*Ficus watkinsiana*), emerge through the canopy and dominate the skyline which is often shrouded in mist and cloud.

The temperature is usually several degrees cooler than on the coastal

Above: Mt Nebo area **Opposite:** Northbrook Creek

Near Wivenhoe Outlook

plain below, making it an ideal place to escape the summer heat. A walk through the rainforests at Mt Glorious provides a good introduction to the large number of plant species present and also the complexity of life forms including many types of buttressing, fungi, ferns, vines and epiphytes.

To the west and also to a lesser extent, the east of Mt Glorious, isolated rocky outcrops and gorges have given rise to small patches of Paperbark Tea-tree (*Melaleuca quinquenervia*) and grass

tree (*Xanthorrhoea latifolia*) dominated heathlands growing amongst the rocks, ledges and scree slopes. These areas are particularly suited to orchids and wildflowers which makes a visit in late winter or spring worthwhile.

The Park Fauna

The most obvious forms of wildlife in Brisbane Forest Park are the birds. With more than 240 species on record, it is one of the best single locations for birdwatching in the country. The rainforest species are spectacular but secretive and it is necessary to walk quietly along a mountain track to catch glimpses of birds such as Noisy Pittas (*Pitta versicolor*), Southern Log Runners (*Orthonyx temminckii*), Paradise Riflebirds (*Ptiloris paradiseus*) and Regent Bowerbirds (*Sericulus chrysocephalus*).

Access to the large, man-made wetland areas in Brisbane Forest Park is restricted for water supply protection purposes, but it is possible to join one of the park's "Go Bush" activities which occasionally visit these areas. The birdlife encountered on these walks, even within a few kilometres of the city, is abundant and

Bush Rat *Rattus fuscipes*

includes White-bellied Sea Eagles (*Haliaeetus leucogaster*), Jacanas (*Irediparra gallinacea*) and Cotton Pygmy -Geese (*Nettapus coromandelianus*).

Many rare and rarely seen bird species are also known to frequent the park and some recent observations have included Red Goshawks (*Erythrotriorchis radiatus*), Marbled Frogmouths (*Podargus ocellatus*), Bush-hens (*Amaurornis olivacea*) and Black-breasted Button Quail (*Turnix melanogaster*). Even the casual visitor to one of the many recreation areas is likely to see Brush-Turkeys (*Alectura lathami*), Satin Bowerbirds (*Ptilinorhynchus violaceus*), Kookaburras (*Dacelo novaeguineae*) and Pied Currawongs (*Strepera graculina*).

The walking tracks in the Mt Nebo area have a reputation for being good bird-watching locations due to their mix of open forest and rainforest which attracts an abundance of species.

Some 66 mammal species have been recorded from the park but most are unlikely to be seen by the casual day visitor. An occasional rainforest pademelon (*Thylogale* spp.) or perhaps even an echidna (*Tachyglossus aculeatus*) shuffling along, are the most

Near Tenison Woods Mountain

Above: Black Tea-trees (*Melaleuca bracteata*)

anyone can expect. However, at night the situation changes and the majority of the park's mammals, which are nocturnal, become active. Red-necked Pademelons (*Thylogale thetis*), Mountain Brushtail Possums (*Trichosurus caninus*) and Common Ringtail Possums (*Pseudocheirus peregrinus*) are commonly encountered on an evening spotlight walk.

In the larger trees, glimpses of several species of gliding possums are possible, whilst on the ground bandicoots (*Isoodon macrourus/ Perameles nasuta*), Long-nosed Potoroos (*Potorous tridactylus*) and several native rats and mice can be found.

Insectivorous bats and the larger, noisier fruit bats are found throughout the park although in the dark a fleeting glimpse through the beam of a spotlight is generally the most that can be expected.

Visitors are unlikely to see koalas and kangaroos in Brisbane Forest Park. These animals are found in small numbers and the probability of encountering one in such a large area is quite low. Platypus (*Ornithorhynchus anatinus*) likewise are secretive and frequent the more remote waterways. Small numbers of dingoes (*Canis lupus dingo*) roam over much of the D'Aguilar Ranges and can sometimes be heard at night.

Reptiles are a conspicuous and active group during the warmer months, the most obvious being the Lace Monitor (*Varanus varius*) or Tree Goanna. The Land Mullet (*Egernia major*), a large, shiny black skink that glides swiftly through the rainforest undergrowth can be seen on warmer days. Less frequently encountered is the Southern Angle-headed Dragon (*Hypsilurus spinipes*), perfectly camouflaged in the dappled light and lichen encrusted environment of buttressed roots and boulders.

Burton's Snake Lizard (*Lialis burtonis*) may at first glance resemble a snake, however this animal is harmless and may drop its tail if harassed. Many snakes are also harmless but can be easily confused with one another. For example the non-venomous Keelback (*Tropidonophis mairii*) is similar to the venomous Rough-scaled Snake (*Tropidechis carinatus*). Both are abundant along the creeks in the park. The two most commonly encountered snakes are the harmless Common Tree Snake (*Dendrelaphis punctulatus*), and the park's largest snake, the Carpet Python (*Morelia spilota variegata*), which grows up to 3 m. Although non-venomous, Carpet Pythons can bite savagely if disturbed.

During the summer, many snakes become nocturnal, and particularly on wet humid nights, can be spotted crossing roads where many become traffic victims. They range from the boldly marked Bandy Bandy (*Vermicella annulata*) to the sinuous, bulbous eyed Brown Tree Snake (*Boiga irregularis*). Small and isolated populations of some of the world's most

Above: Carpet Python *Morelia spilota variegata*

potentially dangerous snakes are also known in the park, and include the Death Adder (*Acanthophis antarcticus*), found on drier ridge tops, and the Taipan (*Oxyuranus scutellatus*) which has recently been recorded from the Northbrook Valley.

Several species of turtles are found in the reservoirs and larger water holes along the creeks of Brisbane Forest Park. Some 26 species of frogs have also been recorded but unfortunately the most common amphibian is the introduced Cane Toad (*Bufo marinus*). The most unusual frog, the Southern Dayfrog (*Taudactylus diurnis*), has not been seen for more than 15 years and can probably be considered extinct. It was an inhabitant of the swift flowing, rainforest streams where it was active on summer days.

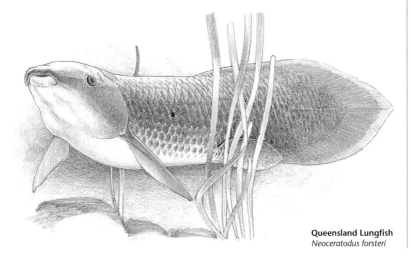

Queensland Lungfish
Neoceratodus forsteri

Northbrook area **Inset:** Green Treefrog (*Litoria caerulea*)

Large tadpoles which can be seen throughout the year in leafy, rainforest pools belong to the Great Barred-frog (*Mixophyes fasciolatus*), the adults of which, when seen on wet summer nights in the rainforest, are strikingly marked and of impressive size. Another rainforest inhabitant is the Mt Glorious Spiny Crayfish (*Euastacus setosus*). This crustacean is endemic to the park and found only in high altitude streams. It is similar and related to the Blue Spiny Cray (*Euastacus sulcatus*) of the McPherson Ranges but lacks the intense blue colouring.

Another notable aquatic animal is the Lungfish (*Neoceratodus forsteri*), unique to South-east Queensland. The lungfish was introduced to Enoggera Reservoir last century for safe-keeping by concerned early conservationists. The most conspicuous fish in the park's creeks and waterholes is the Long-finned Eel (*Anguilla reinhardtii*) because of its unique ability to travel short distances overland (see p. 161). Approximately 15 fish species have been identified in Brisbane Forest Park. — **Martin Fingland and David Morgans**

BAYSIDE REGIONAL PARK

The Bayside Regional Park consists of a network of smaller reserves scattered over 10 km, stretching from Capalaba West to Wynnum North.

The bushland components of this Brisbane City Council Reserve cover some 400 ha and are a haven from the surrounding suburban development. They are bounded by the western bank of Tingalpa Creek from just below the Leslie Harrison Dam to the mouth of the creek and include Lota Creek.

The dry eucalypt forest common to this part of the region covers most of the park with some paperbark forest and mangroves fringing the lower reaches of the creek. A canoe trip from the mouth of the creek upstream provides the best opportunity to see some of Brisbane's typical wildlife — wallabies, birds of prey, wading birds and reptiles. — **Stephen Poole**

Look out for: Carpet Pythons (*Morelia spilota variegata*) coiling on overhanging branches; Bearded Dragons (*Pogona barbata*) sunning themselves.

Brisbane Short-necked Turtle *Emydura signata*

Tingalpa Creek

BOONDALL WETLANDS

Opposite: Aerial view of Nudgee Creek **Above:** Mangroves, Nudgee Beach

The Boondall Wetlands are the largest remaining habitat of their kind within Brisbane City and are internationally recognised as an important feeding and resting place for migratory wading birds.

The wetlands are a Brisbane City Council Reserve covering some 665 ha. This will be almost doubled to about 1200 ha in July 1997 when an additional 540 ha from the Federal Government (Brisbane Airport) is added to the reserve.

Boondall forms part of a series of unconnected wetlands, which include Deagon and Tinchi Tamba (see pp. 65, 75), and which are the remnants of a much larger habitat that existed before European settlement.

Although a wetland, Boondall's vegetation ranges from intertidal mudflats and mangroves to paperbark and dry eucalypt forests. Patches of Forest She-oak (*Allocasuarina torulosa*) and Swamp She-Oak (*Casuarina glauca*)

LOCALITY GUIDE

Location: 15 km north-east of Brisbane GPO; 15 mins' driving.

Access: Car, bike, train or boat. (From Boondall Railway Station use the walkway under the Gateway Arterial Road).

Facilities: Graded walking tracks; bikeways; boat ramps; visitor centre.

Restrictions: No domestic animals; no camping; no motorised vehicles (except boats).

grow near and within the dry eucalypt forest and on the edges of the paperbark forest. Small creeks flowing through the wetlands complete a complex and vital habitat network for birds, fish, reptiles and other wildlife.

The mangrove species at Boondall are typical of those elsewhere in the Greater Brisbane Region and include the Grey (*Avicennia marina*), Red (*Rhizophora stylosa*) and River (*Aegiceras corniculatum*) Mangroves. Within the mangrove forests, the long shiny green fronds of the Mangrove Fern (*Acrostichum speciosum*) are most noticeable on the landward side. This plant is not as common in the region as it once was because of wetland clearance.

The locally rare Pale Sundew (*Drosera peltata*)[1] is an "insect-eating" species also found at Boondall. It can be easily identified by its pink to red leaves which are coated with a sticky substance to catch insects. These tiny plants — about 1.5 cm across — are most likely to be seen on damp, exposed sites (see p. 117).

The wetlands are one of the best places in the region for birdwatching, particularly for wading and migratory birds, with 57 families present, many in large numbers.[2] The Brolga (*Grus rubicundus*), Australian Pelican (*Pelecanus conspicillatus*), Osprey (*Pandion haliaetus*), Peregrine Falcon (*Falco peregrinus*), Wood Duck (*Chenonetta jubata*), Night Heron (*Nycticorax caledonicus*), Bush Stone-curlew (*Burhinus grallarius*), White Winged Tern (*Chlidonias leucopterus*), Bar-tailed Godwit (*Limosa lapponica*), Red-necked Avocet (*Recurvirostra novaehollandiae*),

The Boondall Wetlands are a bird-watcher's haven. Migratory birds from countries such as Siberia, China, Japan, Mongolia and Alaska can be seen from early October to late March. Local species can be spotted all year, particularly in winter. The best viewing spots are the Nudgee Beach foreshore; the area north of the car park at the Boondall roundabout; the Nundah Creek bird hide; and the Nudgee Beach bird hide.

Striated Heron *Butorides striatus*

Lesser Sand Plover (*Charadrius mongolus*), Collared Kingfisher (*Halcyon chloris*), Red-rumped Parrot (*Psephotus haematonotus*), and White-breasted Woodswallow (*Artamus leucorynchus*) are some that have been recorded.

Opposite: Sandflats, Boondall
Below: Mudflats, Cabbage Tree Creek

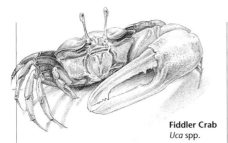

Fiddler Crab
Uca spp.

Boondall is also home to the regionally rare Grass Owl (*Tyto capensis*), which makes a large grass nest on the ground. It is unusual to find Grass Owls so close to a city because most suitable habitats in near urban areas have been cleared or disturbed.

Boondall is a favoured location for frog enthusiasts. The Broad-palmed Rocket Frog (*Litoria latopalmata*), Naked Tree-frog (*Litoria rubella*), Ornate Burrowing Frog (*Limnodynastes ornatus*) and the Beeping Froglet (*Crinia signifera*) have all been recorded there.[3] As for most places, the best times to see or listen to frogs are summer evenings after rain.

One of the most unusual animals at Boondall is the Tree Skink (*Egernia striolata*) which has previously not been recorded east of the Great Dividing Range in southern Queensland.[4] Why this predominantly dark grey skink, which can grow up to 30 cm, occurs here is a biological enigma, because it appears to be so far out of place. This one example reinforces the biological richness of the Greater Brisbane Region.

Some intertidal areas at Boondall, are listed under the Ramsar Convention on Wetlands of International Importance. Boondall is also covered by the JAMBA and CAMBA — Japan/ Australian and China/Australia migratory bird agreements. These conventions require all levels of Australian government to protect the birds, their habitats and undertake or encourage research programs. — **Stephen Poole**

Look out for: Marine snails on the mudflats; and Illidge's Ant Blue Butterfly (*Acrodipsas illidgei*), a species restricted to mangroves.

The Environmental Education Centre run by the Queensland Department of Education at Nudgee Beach has been undertaking base studies of the ecology of Boondall. Particular emphasis has been placed on marine invertebrate species of the intertidal mudflats.

The Brisbane City Council's Visitor Centre is next to the Gateway Arterial Road, opposite the Entertainment Centre.

Above: Nundah Creek **Opposite:** Nudgee Beach

BRISBANE KOALA PARK

The Brisbane Koala Park is a series of conservation reserves which stretch from Mt Petrie to Ford Road at Burbank and along Tingalpa Creek on Brisbane's southside. Most of these sites are already linked and it is expected that when land acquisitions are completed, this park will be the largest of its type in Australia.

The koala population to the south-east of Brisbane, estimated at 3000–5000 animals, is one of Australia's largest and most significant. It attests to the importance of the lowland eucalypt forests in southern Queensland.

In the Mt Petrie section of the park, the extensive dry eucalypt forests and woodlands are dominated by Scribbly Gum (*Eucalyptus racemosa*). There is also some wallum heath where the distinctive Broad-leaved Banksia (*Banksia robur*) grows. The leaves of this plant are up to 30 cm long, have serrated edges and an almost rubbery texture.

At the Tingalpa end of the park, the forest composition changes to include Grey Gum (*E. propinqua*), Spotted Gum (*Corymbia variegata*), Brush Box (*Lophostomen confertus*), and Tallow-wood (*E. microcorys*).

The park also has a small remote patch of rainforest at the junction of Tingalpa and Buhot Creeks.

Although the park was primarily established to protect koalas, other animals benefit too. For example, the Greater Glider (*Petauroides volans*), which is threatened in the Brisbane region, and Squirrel Glider (*Petaurus norfolcensis*) have been observed here.
— **Stephen Poole**

Opposite: Burbank area **Inset:** Broad-leaved Banksia *Banksia robur*

L O C A L I T Y   G U I D E

Location: 14–20 km south-east of Brisbane GPO; 25 mins' driving.

Access: Via Mt Cotton and Mt Gravatt-Capalaba Roads; numerous other entry points.

Facilities: Rough walking tracks through most reserves, some graded tracks.

Restrictions: No domestic animals; no camping; no motorised vehicles.

Most people think that koalas will eat the leaves of only a handful of eucalypts but they are known to feed on more than 20 species. In local areas, this may decrease to three or four preferred trees such as the Grey Gum, Tallowwood, and Forest Red Gum. At certain times of the year, Brush Box, which is not a eucalypt, may also become a favourite food tree. Koalas also occasionally eat the leaves of Paperbark Tea-trees. — SP

The Brisbane regional koala habitat (see map below) covers about 25,000 ha of south-eastern Brisbane and overlaps the boundaries of Logan City and Redland Shire. The area contains one of Australia's largest koala populations. Much of the bushland is privately owned, however a number of reserves have been set aside to protect core habitat areas. The Brisbane Koala Park, Venman's Bushland National Park, Daisy Hill State Forest and several Redland Shire Council conservation reserves are within the "boundaries" of the habitat.

Daisy Hill Koala Centre (see p. 64) gives visitors the chance to see koalas up close and learn about these appealing Australian marsupials. Interactive displays throughout the centre explore how koalas live and survive. The Midnight Woodland Theatre invites visitors to discover the bush at night. Rangers and volunteers are on hand to answer questions on conserving koalas and their habitat. The Koala Centre is open from 10 am – 4 pm seven days a week. Telephone for opening hours on public holidays. (Wheelchair access.)

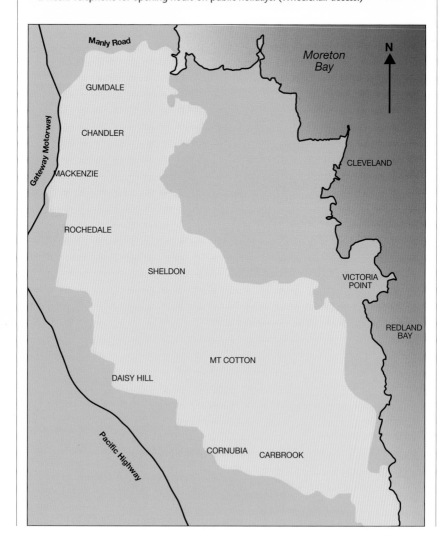

DAISY HILL STATE FOREST PARK

At 435 ha, Daisy Hill State Forest is one of the smallest state forests in the Greater Brisbane Region, yet it has a history which vividly illustrates the community's changing perceptions regarding the value of natural areas.

When the first ranger visited Daisy Hill in 1903 he found a forest that was being devastated by the illegal removal of timber. (The area was reserved for timber use early this century.) By 1930, a forest ranger was permanently stationed at Daisy Hill. He lived in a tent and his job was to protect the forest from fire, illegal cattle grazing and timber theft.

Throughout the 1930s there was increasing community pressure to use the area for grazing, banana growing, mining and settlement. (Evidence of early gold mining — the collapsed and

sealed entrance of a gold mine — can still be seen near the top picnic area.) However these ventures were resisted and gradually the forest became more important to the local community for its natural values.

Ridgeline forest

The topography of Daisy Hill State Forest is mostly gently undulating but it includes some steep areas, particularly on the Springwood-Shailer Park escarpment.

The vegetation is predominantly dry eucalypt forest and woodland. Paperbark tea-trees (*Melaleuca quinquenervia*) grow along the Buhot Creek lagoon system which runs through the centre of the park.

The forests of Daisy Hill contain Broad-leaved and Grey Ironbarks (*Eucalyptus fibrosa* and *E. siderophloia*) and Spotted Gum (*Corymbia variegata*) — typically on the ridges and upper slopes.

In lower areas, visitors are more likely to see Tallowwood (*E. microcorys*) and Grey Gum (*E. propinqua*) which sheds its pale outer bark to reveal fresh orange bark. Other forest species include Broad-leaved White Mahogany (*E. carnea*) and in the north-west of the forest, Gum-topped Box (*E. moluccanna*). Scribbly Gum (*E. racemosa*) is found in the woodland areas.

The small and prickly *Daviesa villefera* can be found amongst the understorey species at Daisy Hill which can vary from dense acacia scrub through grassy areas with the odd banksia to dry rainforest species and heathland.

The eucalypt communities of Daisy Hill provide habitats for many of Brisbane's native animals. The area is well represented with birds, mammals, reptiles, amphibians, fish and crustaceans. Visitors who take the time to explore are likely to hear, and perhaps see, the Eastern Whipbird (*Psophodes olivaceus*) in the wetter parts of the park or Red-backed Fairy-wrens (*Malurus melanocephalus*). Two of the most common birds are the Grey Fantail (*Rhipidura fuliginosa*) and Yellow-faced Honeyeater (*Lichenostomus chrysops*).

During the day, park visitors may be lucky enough to see one of the forest's two wallaby species — the Swamp Wallaby (*Wallabia bicolor*) with its black tail or the Red-necked Wallaby (*Macropus rufogriseus*).

The Buhot Creek lagoon system supports a diverse range of frogs, fish, shrimp and crayfish and is the only known mainland location, in the Greater Brisbane Region, of the Soft-spined Sunfish (*Rhadinocentris ornatus*). These aquatic environments are important remnants of a previously much wider distribution of freshwater habitats. The Great Barred-frog (*Mixophyes fasciolatus*) can be found on the southern slopes of the forest.

Opposite: Buhot Creek **Above:** Dry eucalypt forest

Buhot Creek

During summer months bushwalkers might notice what appear to be small grey rocks on the walking trails. This is in fact the Mountain Katydid (*Acripeza reticulata*). The flightless female has a spectacular defence — when disturbed its wings fold back to reveal an orange collar and bright blue and red stripes on the abdomen.

Daisy Hill is also home to one of Australia's most recognisable animals — the Koala. Visitors can learn more about the Koala (*Phascolarctos cinereus*) at the Department of Environment's Koala Information Centre in the central picnic area (see p. 60).

Daisy Hill is one of the most heavily used state forests in Queensland and a favourite location for orienteering, horse-riding and mountain bike riding. Almost 250,000 people visit the park annually and this imposes a new kind of pressure. The challenge now is to ensure that these newer recreational demands are managed in a way that will allow the integrity of the park to be maintained. — **Mark Peacock**

Look out for: Yellow-faced Whip Snakes (*Demansia psammophis*) rustling at the side of the track; goannas, Lace Monitors (*Varanus varius*), crashing through the undergrowth.

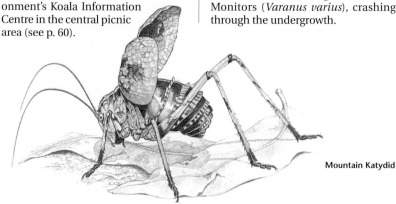

Mountain Katydid

DEAGON WETLANDS

The Deagon Wetlands, which cover 50 ha, form the largest single remaining area of Paperbark Tea-tree (*Melaleuca quinquenervia*) forest in Brisbane City. They are located just north of the Deagon Racecourse and are bounded on the western edge by the Gateway Arterial Road.

Paperbark forest has the highest loss rate and is under the most threat of any vegetation type in South-east Queensland. This, and its relatively undisturbed nature, make the Deagon Wetlands one of the most important bushland sites within the metropolitan area. The wetlands are administered by Brisbane City Council as a Conservation Reserve, specifically established to protect this fast disappearing habitat.

Apart from the dominant paperbarks, eucalypts such as the Grey Gum (*Eucalyptus propinqua*) and Narrow-leaved Ironbark (*E. crebra*) are scattered through the wetlands along with Brush Box (*Lophostemon confertus*) and Red Ash (*Alphitonia excelsa*). Red Ash can be distinguished by its leaves which are dark green with a rubbery surface on top and very pale or whitish on the underside. The forest understorey is mostly a carpet of Bungwall Fern (*Blechnum indicum*).

During dry periods, particularly at the end of winter, there are virtually no wet areas within the forest. Water marks on the trees indicate the extent to which summer rainfall inundates the area. A water table that is close to the surface enables the paperbarks to continue growing during dry periods.

Although relatively small, Deagon supports a broad range of fauna species — many of which are also found in Boondall and Tinchi Tamba (see pp. 53, 75). Due to its size, Deagon

LOCALITY GUIDE

Location: 17 km north-east of Brisbane GPO; 20 mins' driving.

Access: Via Bracken Ridge Road or Kempster Street, Sandgate.

Facilities: None.

Restrictions: No domestic animals; no camping; no motorised vehicles.

Bungwall Fern, foreground

Opposite: Paperbark forest **Inset:** Smooth-barked Apple (*Angophora leiocarpa*) **Above:** After rain

does not have any larger mammals such as wallabies or kangaroos but it does have an abundance of reptiles and amphibians.

A quiet walk through the reserve might reveal the Long-necked Turtle (*Chelodina longicollis*), a common reptile of the Greater Brisbane Region, which secretes a pungent musk odour when disturbed or handled. It is frequently seen crossing roads after heavy rain. (Greater Brisbane contains 4 of Australia's 15 species of turtles).

Burton's Snake Lizard (*Lialis burtonis*) which grows to about 30 cm may also be encountered. It is frequently misidentified as a snake due to its pointed head and snout. Verreaux's Skink (*Anomalopus verreauxii*) resembles a long, thick worm about 1 cm in diameter and up to 30 cm long. It is the largest burrowing skink in the region and tends to avoid sunlight. It is most likely to be seen above ground after heavy rain has flooded its burrows.

As in most wetlands, a late afternoon or evening of "frogging" can be rewarded with sightings or the calls of the Striped Marshfrog (*Limnodynastes peronii*), Graceful Treefrog (*Litoria gracilenta*), Striped Rocketfrog (*L. nasuta*) and Eastern Sedgefrog (*L. fallax*).

Even though wetlands are fragile ecosystems highly susceptible to human impacts, the presence of these frogs indicates that Deagon is a viable environment. Frogs are regarded as accurate biological indicators of environmental damage.

Bushwalkers in the wetlands are often surprised by a whirring and flapping blur of feathers. The culprit is the Brown Quail (*Coturnix australis*) which frequents dense undergrowth and is easily startled. — **Stephen Poole**

Look out for: Lots of birds including the Sacred Kingfisher (*Halcyon sancta*), Clamorous Reed Warbler (*Acrocephalus stentoreus*) and Golden-headed Cisticola (*Cisticola exilis*).[5]

Striped Rocketfrog

INDOOROOPILLY ISLAND

Courtesy of Dr L. Hall.

Indooroopilly Island, only a few kilometres from the heart of Brisbane, is a mangrove island and most noticeable as an area of natural vegetation in heavily urbanised surroundings.

Species include the Grey (*Avicennia marina*) and River (*Aegiceras corniculatum*) Mangroves. A small area of eucalypt forest on the island is dominated by Forest Red Gum (*Eucalyptus tereticornis*).

The island is a Department of Environment Conservation Park and is listed on the Register of the National Estate because of the large colonies of Grey-headed and Black Flying Foxes (*Pteropus poliocephalus* and *P. alecto*), which use the island as a breeding site. The Little Red Flying Fox (*P. scapulatus*) has also been recorded.[6] (This means that the island is used by three of Australia's four species of flying foxes.)

While flying foxes can be seen at most times of the year, it is during the summer months that up to several

LOCALITY GUIDE

Location: 7 km west of Brisbane GPO.

Access: Boat

Restrictions: No domestic animals.

hundred thousand bats roost on the island. The best time to see the bats is at dusk, when they fly off in search of blossoms and fruit — a truly unforgettable sight.

The importance of Indooroopilly Island increases as other flying fox colonies in South-east Queensland are disturbed or destroyed.

The "Batty Boat" tours organised by the Wildlife Preservation Society of Queensland offer opportunities to observe the dusk fly-off from the water (see Useful Contacts p. 208). Alternatively, canoeing or viewing from the opposite bank can be equally rewarding, particularly with a good sunset as a backdrop. — **Stephen Poole**

KARAWATHA FOREST RESERVE

Karawatha Forest is a Brisbane City Council Reserve covering 900 ha. It is important not only as a large bushland habitat, but also as an open space buffer between residential areas on Brisbane's south-side and the northern boundary of Logan City. This is particularly so given the continuing rapid loss of bushland in South-east Queensland.

Karawatha is part of the coastal lowlands but there are some higher areas of rocky sandstone outcrops where the headwaters of Scrubby, Slacks and Bulimba Creeks form.

Dry eucalypt forest grows on the slopes and lowlands of Karawatha with woodland on the ridges. Both have heath understoreys. A small wallum heath within the reserve is characterised by the Broad-leaved Banksia (*Banksia robur*), which is common to the Greater Brisbane Region, and grass trees such as *Xanthorrhoea macronema*. Pockets of Paperbark Tea-trees (*Melaleuca quinquenervia*) occur in wetter areas around creeks and the Illaweena Lagoons. These waterways

LOCALITY GUIDE

Location: 18 km south of Brisbane GPO; 30 mins' driving.

Access: Rail access via Trinder Park Station; via Compton and Acacia Roads, Kuraby; and Illaweena Street picnic area, Drewvale.

Facilities: Picnic areas; bush tracks.

Restrictions: No domestic animals; no camping; no motorised vehicles.

are one of the few remaining examples of the lagoon systems that would have existed across the region at the time of European settlement.

Above: Common Dunnart *Sminthopsis murina*
Below: Illaweena Lagoons

Above: Illaweena Lagoons

Karawatha contains 324 plant species.[7] Three locally rare and restricted trees grow on the sandstone ridges in the reserve — Planchon's Stringybark (*Eucalyptus planchoniana*), Bailey's Stringybark (*E. baileyana*) and Plunkett Mallee (*E. curtisii*).

Planchon's Stringybark, also known as Needlebark, has an orange-brown bark and is easily identified by its very large and distinctively shaped seed capsules (gumnuts) which resemble a giant goblet.

As with many bushland areas around Brisbane, some habitat degradation has occurred due to invasions by pest species along former four-wheel-drive and trail bike tracks and associated erosion problems. Rubbish dumping in parks and reserves around the Greater Brisbane Region remains a problem for local authorities and state government agencies alike.

However the diversity of habitats within the park — heath to freshwater wetlands — provides for an abundance of wildlife. Karawatha is particularly noteworthy because it still has kangaroos and wallabies — animals which have disappeared from many other smaller, natural areas. There are populations of Grey Kangaroos (*Macropus giganteus*) and Red-necked Wallabies (*M. rufogriseus*).

Two tiny marsupials, the Common Dunnart (*Sminthopsis murina*) and Common Planigale (*Planigale maculata*) have been recorded in Karawatha along with 11 species of equally small insectivorous bats (micro-bats).[8] The Dunnart and Planigale are often referred to as "marsupial mice" and, to the untrained eye, could easily be mistaken for the introduced House Mouse (*Mus musculus*).

The forest also still has native rodents such as the Swamp Rat (*Rattus lutreolus*) which, like the kangaroos and wallabies and many other ground animals, has disappeared from most other urban reserves (see Toohey Forest p. 79)

Brisbane is the only major Australian city to still have platypus and they have been observed in Scrubby Creek and the Illaweena Lagoons. Platypus burrows can extend for up to 15 m away from the creek bank.

Damage to the burrow systems, along with channelisation (concreting) of creeks and water pollution, are some of the reasons why platypus have been lost elsewhere. This emphasises the importance of maintaining creek vegetation buffers.

Karawatha is increasingly being appreciated as an amphibian habitat. Two species of frogs considered rare and vulnerable — the Green-thighed Frog (*Litoria brevipalmata*) and the Wallum Froglet (*Crinia tinnula*) — are among the 18 species recorded at Karawatha. These were only recently discovered and with further investigation it is likely other significant amphibian species will be found.

More than 100 native bird species are known from the reserve including the locally rare Glossy Black Cockatoo (*Calyptorhynchus lathami*) which can sometimes be seen feeding in casuarina trees.

Casuarinas, usually considered a prolific pioneer species and sometimes a nuisance, are a major food source for the cockatoos which break open and eat the seed capsules.

The Karawatha Forest Protection Society organises bush walking and nature trips within the park (see Useful Contacts p. 208). — **Stephen Poole**

Look out for: Squirrel Gliders (*Petaurus norfolcensis*) when spotlighting; inquisitive Kookaburras (*Dacelo novaeguinea*) in low branches beside the tracks.

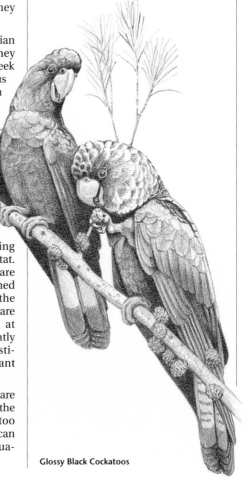

Glossy Black Cockatoos

TINCHI TAMBA

Tinchi Tamba, a wetland covering 371 ha, is located on the flood plain of the Pine River and most of the area experiences regular inundation.

This Brisbane City Council Reserve is important because it encompasses so many habitat types — dry eucalypt and paperbark forests, casuarina forests, mangroves, claypans and salt-marshes, grasslands, swamps, some cleared areas and former pasture.

The reserve, originally part of a grazing lease, was intended to become a canal estate. Fortunately this did not happen and parts of Tinchi Tamba — those visited by migratory wading birds — have now been included in the Moreton Bay Marine Park and, like Boondall, are listed on the Ramsar Convention Register of Wetlands of International Importance (see p. 56).

Following the cessation of grazing in the early 1990s, the vegetation communities are slowly recovering and

Hermit Crab *Clibanarius taeniatus*

Opposite and above: Bald Hills Creek

Saltmarsh

some like the mangroves will gradually increase. In the long term this will make the reserve more attractive to wildlife.

Nearly all the mangrove species of the Greater Brisbane Region can be found within Tinchi Tamba. The Grey Mangrove (*Avicenna marina*) is most common but the community also contains the Orange (*Bruguiera gymnorhiza*), Milky (*Excoecaria agallocha*), Red (*Rhizophora stylosa*), River (*Aegiceras corniculatum*) and rare Black (*Lumnitsera racemosa*) Mangroves.

In the dry eucalypt forests, the main species are Forest Red Gum (*Eucalyptus tereticornis*), Grey Ironbark (*E. siderophloia*) and Moreton Bay Ash (*Corymbia tessellaris*). Other species include *Brachychiton populneus* and the Rock Fig (*Ficus platypoda*).

More than 200 species of birds are known from Tinchi Tamba and this highlights its significance as a bird-watching site. It is not only the number of species which is impressive but also the number of individual birds. For example, Magpie Geese (*Anseranas semipalmata*) are usually only seen in small numbers, but at Tinchi Tamba groups of more than 70 birds have been recorded.

Tinchi Tamba is one of the few places in the Greater Brisbane Region where Ospreys (*Pandion haliaetus*) nest. Ospreys are becoming uncommon in urban and near-urban areas due to the removal of suitable habitat — tall, dead trees that afford good views of their surrounds to guard against predators and watch for food. The reserve is also the site of one of the largest Cattle Egret (*Ardea ibis*) rookeries in the region. (The largest is near Gatton).

Most of the wading birds, including whimbrels, knots and sandpipers can be seen on the mudflats at low tide, particularly in the bend of Bald Hills Creek in summer. The woodlands contain everything from cuckoos and kingfishers to larks and honeyeaters.

As yet, little information about mammal species in the reserve is available. This is because fauna survey work has not been undertaken and because the effects of predation by domestic animals from the surrounding residential development are not clear.

Tinchi Tamba probably has Squirrel, Sugar and Feathertail Gliders (*Petaurus norfolcensis*, *P. breviceps* and *Acrobates pygmaeus*). The Feathertail Glider is more common across Brisbane than

Brahminy Kite

most people realise as the small size of the animal makes it difficult to locate, especially when spotlighting.

Tinchi Tamba, like Boondall and Deagon, is an ideal habitat for frogs. Species such as the Clicking Froglet (*Crinia parinsignifera*) and the Striped Rocketfrog (*Litoria nasuta*) can be found there along with the Spotted Marshfrog (*Limnodynastes tasmaniensis*) and Ornate Burrowing-frog (*L. ornatus*).

Marine invertebrates (worms, pippis, small crustaceans and the like) are poorly known but it is likely that with intensive surveying, many species new to the region will be discovered as is occurring at Boondall.[11] One survey has revealed 22 species of crabs in estuarine areas across Tinchi Tamba.[12]

Look out for: Whistling Kites (*Haliastur sphenurus*) and Brahminy Kites (*H. indus*) soaring above the waterways; masses of soldier crabs scuttling through mangrove forests. — **Stephen Poole**

TOOHEY FOREST

Toohey Forest stretches through the inner southern suburbs of Brisbane from Mt Gravatt Lookout to Tarragindi. The forest covers some 650 ha and most of this bushland is included in the Brisbane City Council Reserve, the Tarragindi Environmental Park and the Griffith University campus.

Fraser's Land Snail
Sphaerospira fraseri

Before the forest was recognised for its environmental values, a number of activities encroached on its periphery. These include the Griffith University campuses of Mt Gravatt and Nathan; the QE2 Sports Complex; Department of Business, Industry and Regional Development Research Park; Mt Gravatt Cemetery; and some freehold (private) bushland.

Toohey has been isolated from other natural areas by urban development and the major roads which encircle and pass through the forest. This, combined with intensive visitor usage, means that the forest has been damaged.

For example, there is evidence of eucalypt dieback near the edges of the South-east Freeway which dissects the forest. The freeway has altered rainfall run-off patterns and exposed more of the forest to the effects of weather — storms in particular.

Ironically perhaps, Toohey's proximity to Griffith University makes it one of the most studied bushland areas anywhere. Toohey contains 450 species of plants including 18 types of eucalypts[13] and 30 different ferns.[14]

Blue-tongued Skink *Tiliqua scincoides*

Most of the forest is dry eucalypt with a shrubby and heath understorey — a habitat type which is declining in South-east Queensland. Dominant trees include Grey Ironbark (*E. sidero-phloia*), Smudgee (*Angophora wood-siana*), Queensland White Stringybark (*E. tindaliae*), Brush Box (*Lopho-stemon confertus*) and Spotted Gum (*Corymbia variegata*). Hairy Bush Pea (*Pultenaea villosa*) and the prickly *Daviesa villefera* are common in the understorey.

The wetter riverine areas such as the headwaters of Mimosa Creek provide a refuge from the dry conditions. Many native vines and creepers and some rainforest species grow in the moist conditions.

As in Karawatha, the higher areas of Toohey are sandstone and also have stands of Planchon's Stringybark (*E. planchoniana*) and Bailey's Stringy-bark (*E. baileyana*) — see p. 20.

The impact of "edge effects" — isolation, road traffic and predation by cats and dogs and the like — has affected not only the condition of the vegetation but also the forest's fauna diversity (see above). There may no longer be any native ground mammals in Toohey.

The forest also has nesting Powerful Owls (*Ninox strenua*), a locally rare species which requires a large feeding area and which previously had been observed only as a "vagrant". More than 90 species of birds have been recorded in the forest.

The Blue-tongued Skink (*Tiliqua scincoides*) is found in Toohey as it is

"Edge effects" are external factors which either individually, or in combination, can affect the quality of natural habitats. They include fragmentation and isolation due to surrounding or encroaching development; the introduction of weeds and diseases; changes in watertables or rainfall run-off; pollution; exposure to weather; increased fire rates; loss of hollow-bearing trees; predation by domestic animals; reduced territories and altered diets; increased pest attack; erosion and overuse; and the introduction of other species. Edge effects can devastate the vegetation and fauna populations of an area — as is believed to have occurred in Toohey Forest.

in reserves and backyards across the Greater Brisbane Region. This innocuous but beautiful lizard is so common it is often ignored and because it is so slow-moving is frequently attacked by domestic pets or killed by cars.

Toohey Forest is used by orienteering clubs and other recreational/nature groups. Guides to the vegetation and fauna of Toohey Forest have been produced by the Division of Environmental Science at Griffith University and the Toohey Forest Protection Society organises Sunday Walks through the forest (see Useful Contacts p. 208). — **Stephen Poole**

Look out for: Hard-to-see Tawny Frogmouths (*Podargus strigoides*) (look for the tuft of feathers above the beak); and the well known Toohey Forest crows (*Corvus orru*).

VENMAN'S BUSHLAND NATIONAL PARK

Mr Jack Venman gifted this property to the people of Queensland in 1970. He remained on the property as caretaker and would often stop and tell visitors about his bushland reserve. Sadly, Jack passed away in 1994; he was in his eighties.

This park has long been popular with families and school groups and it is one of the best places to see wallabies in the Greater Brisbane Region. Swamp Wallabies (*Wallabia bicolor*) and Red-necked Wallabies (*Macropus rufogriseus*) used to gather around Jack's homestead, (although the Swamp Wallaby normally prefers to stay in wetter, thicker areas of the forest).

The new national park of 405 ha is larger than the original reserve which was previously managed as an Environmental Park by the Redland Shire Council. The headwaters of Tingalpa Creek form in the park's low hills which run north from the Springwood–Shailer Park escarpment.

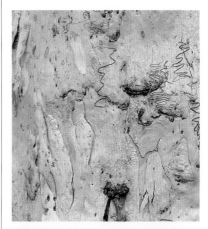

Red-necked Wallaby

Opposite: Upper reaches of Tingalpa Creek
Above Top: Scribbly Gum *Eucalyptus racemosa*
Above: Spotted Gum *Corymbia variegata*

Above : Koala, *Phascolarctos cinereus*
Below: Paperbarks

Most of the park is the lowland dry eucalypt forest typical of the region and the trees are similar to those on other nearby reserves except for the Blackbutt (*Eucalyptus pilularis*). This species has been a favourite timber for logging since European settlement. As with many eucalypts, logging has not endangered the Blackbutt, but it has removed many of the giant examples which once existed across the region.

Other species include Scribbly Gum (*Eucalyptus racemosa*), Tallowwood (*E. microcorys*), Grey Ironbark (*E. siderophloia*), Grey Gum (*E. biturbinata*), Spotted Gum (*Corymbia variegata*)

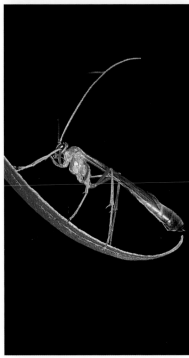

Wasp, *Enicospilus* sp.

and Pink Bloodwood (*C. intermedia*). Brush Box (*Lophostemon confertus*), which has a "box" bark at the base and pink or cream gum (smooth) bark at the top, is also common. A long walk from Venman's through connecting reserves to Daisy Hill State Forest Park and back is a pleasant way to see a variety of eucalypt communities.

Some Paperbark Tea-trees (*Melaleuca quinquenervia*) grow along a tributary of Tingalpa Creek which runs through the park. Significant numbers of the locally rare Eprapah Wattle (*Acacia perangusta*) are protected in the park.

Venman's Bushland National Park falls within the Brisbane regional koala habitat (see p. 60), so watch out for koalas resting or feeding in the tops of trees. Sometimes it can be easier to look for their distinctive droppings (pellets 2–3 cm long and about 1 cm in width) or claw marks (two long scratches, 7–8 cm each, at about 60 deg). — **Stephen Poole**

Look out for: Wrens and fantails in thick undergrowth; and bandicoots.

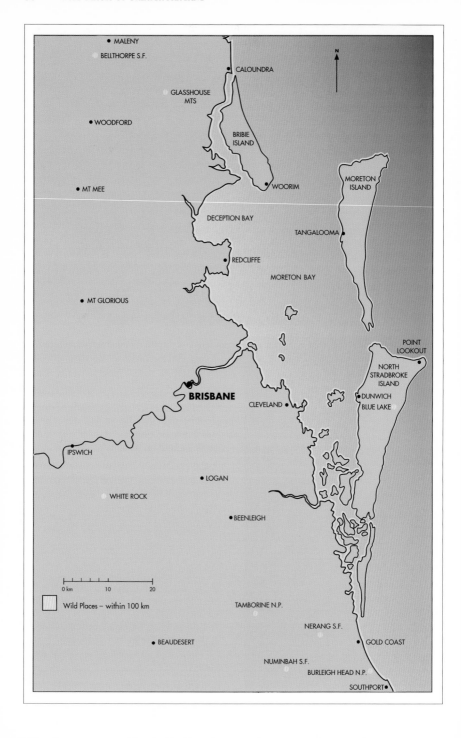

N

MALENY
BELLTHORPE S.F.
CALOUNDRA
GLASSHOUSE
MTS
WOODFORD
BRIBIE
ISLAND
MT MEE
WOORIM
MORETON
ISLAND
DECEPTION BAY
TANGALOOMA
REDCLIFFE
MORETON BAY
MT GLORIOUS
POINT
LOOKOUT
NORTH
STRADBROKE
ISLAND
BRISBANE
DUNWICH
CLEVELAND
BLUE LAKE
IPSWICH
LOGAN
WHITE ROCK
BEENLEIGH
0 km   10   20
Wild Places – within 100 km
TAMBORINE N.P.
NERANG S.F.
BEAUDESERT
GOLD COAST
NUMINBAH S.F.
BURLEIGH HEAD N.P.
SOUTHPORT

WILD PLACES — WITHIN 100 KM

Bribie Island

BELLTHORPE STATE FOREST

Nestled in the foothills of the Conondale Range to the west of Woodford is Bellthorpe State Forest. This forest is a little more difficult to get to and campers need to be self-sufficient because there are no developed areas.

The rugged landscape contains eucalypt forest and rainforest with many small waterfalls and cascades in the creek systems. The forestry roads and old tracks are ideal for exploring the forest and its hidden places.

The mixing of rainforest and eucalypt forest here has resulted in a rich fauna. Cascade Treefrogs (*Litoria pearsoniana*) and Emerald-spotted Treefrogs (*L. peronii*) are just two species of amphibians that can be found in the numerous waterways.

At Bellthorpe, the remains of the old sawmill can still be seen and they are a reminder of past and present uses of the forest. Today, the forest is also managed to provide recreation and conservation benefits and catchment protection.

LOCALITY GUIDE

Location: 95 km north-west of Brisbane GPO; 1 hours' driving.

Access: Take the Caboolture turn-off from the Bruce Highway, follow the D'Aguilar Highway to Woodford to reach Stony Creek, travel 6 km from Woodford and then turn right into Stony Creek Road and follow the signs.

Facilities: Picnic area, toilets, barbecues, water, tables; swimming; self-registration camping; walking tracks; lookouts.

Restrictions: Dogs must remain on a lead; dogs are not permitted in the camping area; 4WD permits required.

Four-wheel-driving is possible along forestry roads however a permit is necessary. Stony Creek day use area is a small picnic ground beside a swimming hole at the junction of Branch and Stony Creeks. — **Toni Hess**

Look out for: Yellow-bellied Gliders (*Petaurus australis*) feeding high in the canopy at dusk; Carpet Pythons (*Morelia variegata spilota*) lying in the winter sun.

Opposite: Sunrise **Above:** Wet eucalypt forest

BRIBIE ISLAND

From a naturalist's perspective, Bribie Island is renowned for two things — its birdlife and its wildflowers.

Bribie is the most northerly sand island in Moreton Bay and is about 33 km long and 8 km at its widest. It differs from Moreton and North Stradbroke Islands by lacking hills or mountains, in fact most of Bribie is only a few metres above sea level.

Bribie's extensive bird habitats and the wallum heathland where wildflowers grow in abundance have benefited from the expansion of protected areas on the island. About one-third of the island's 14,400 ha is now protected through the Bribie Island National Park (4770 ha) and Buckley's Hole Conservation Park (87.7 ha).

The national park forms a narrow band around the island from White Patch on the western side, north to the Spit and south along the eastern shoreline to Freshwater Creek, and then south-west almost to the Woorim/Bongaree Road.

There is an exotic pine plantation in the centre of the island and a 12 km long swamp running north-south

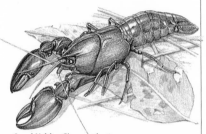

Sand Yabby *Cherax robustus*

through the middle of the plantation. The southern and south-western part of Bribie is mostly residential, commercial and light industrial with the exception of Buckley's Hole Conservation Park.

Pumicestone Passage, and indeed all waters around Bribie Island, are protected by marine reserves. Mainland mangrove forests which fringe the passage are also protected.

Bribie has an interesting range of habitats which include forests dominated by Scribbly Gum (*Eucalyptus racemosa*), Black Sheoak (*Allocasuarina littoralis*) and Cypress Pine (*Callitris columellaris*). Cypress Pine is also commonly called Bribie Island Pine or Coastal White Cypress.

Opposite: Glasshouse Mountains across Pumicestone Passage **Left:** Bribie Island Pines

Pumicestone Passage

The island's dry eucalypt forests usually have grassy understoreys although Bungwall Fern (*Blechnum indicum*) occurs in the understorey of several vegetation communities.

Wallum Banksia (*Banksia aemula*) is common in the low woodland and wallum heath. Wildflower displays can be seen at different times of the year depending on the plants. For instance, the magnificent banksias — including

Broad-leaved Banksia (*Banksia robur*), Red Honeysuckle (*B. serrata*), Dwarf Banksia (*B. oblongifolia*), and Coast Banksia (*B. integrifolia*) — flower through the first half of the year.

Paperbark Tea-tree (*Melaleuca quinquenervia*) forests, interspersed with Swamp Mahogany (*Eucalyptus robusta*) and Forest Red Gum (*E. tereticornis*), grow in wetter areas. Mangroves and tidal wetlands including saltmarsh can be found on the margins of Pumicestone Passage.

A number of animal species that are rare or absent in the Greater Brisbane Region survive on Bribie such as the Brush-tailed Phascogale or Tuan (*Phascogale tapoatafa*). The island still has Emus (*Dromaius novaehollandiae*), although it is uncertain how long they will persist. Buckley's Hole is a favourite haunt of birdwatchers and more than 190 species of birds have been recorded there.[1]

The beautiful Rainbow Bee-eater (*Merops ornatus*) can be seen across Bribie and in the more open forest areas, Variegated Fairy-wrens (*Malurus lamberti*) and the Red-backed Wren (*M. melanocephalus*) are a common sight. The heathlands attract honeyeaters to the callistemons and banksias.

Above: Near Red Beach **Insets:** left, Crested Tern *Sterna bergii*, right, Jellyfish

Water birds include the Australian Pelican (*Pelecanus conspicillatus*), Little Black Cormorant (*Phalacrocorax sulcirostris*), Darter (*Anhinga melanogaster*), Jabiru (*Xenorhynchus asiaticus*), Sacred Ibis (*Threskiornis aethiopica*), Plumed Whistling Duck (*Dendrocygna eytoni*), and Chestnut Teal (*Anas castanea*).

Remember that many wading birds are migratory and are more likely to be seen during the Northern Hemisphere winter. During the season watch out for the Terek Sandpiper (*Xenus cinereus*), Common Sandpiper (*Actitis hypoleuces*), Latham's Snipe (*Gallinago hardwickii*), and the Whimbrel (*Numenius phaeopus*). — **Stephen Poole**

Look out for:
The tracks of birds in mud and on the beach; dugongs (*Dugong dugon*) grazing seagrass beds in Pumicestone Passage.

Bribie Island National Park requires 4WD use or boat/walking access to see much of the park in a reasonable time frame. There are picnic/camping areas at Mission Point, Lime Pocket, Westaways Creek, Lighthouse Reach, Lions Park, Poverty Creek and Gallagher Point. The Ranger's Station is located at White Patch just north-west of the township of Bellara and Banksia Beach.

BURLEIGH HEAD NATIONAL PARK

The Gold Coast, due to its rapidly expanding urban areas, has a scarcity of natural habitats. One that is prominent but tends to be overlooked is Burleigh Head National Park which covers only 27.6 ha.

Although small, this park is popular with schools for field visits. Its proximity to developed areas, small size and heavy usage have caused some habitat degradation. However, the park remains an outstanding natural area because of its diverse vegetation and geological phenomena.

Burleigh Head is a remnant lava flow of the Tweed Shield Volcano centred on Mt Warning in northern NSW, which erupted around 23 million years ago. This same huge volcano also formed the Springbrook, Lamington and Tamborine Plateaus (see pp. 123, 157 and 181). The spectacular, six-sided basalt columns found on the Tugun Lookout

(see pp. 123, 157 and 181)

LOCALITY GUIDE

Location: 100 km south of Brisbane GPO; about 1.5 hours' driving.

Access: Gold Coast Highway via West Burleigh.

Facilities: Picnic area, tables, shelter shed, toilets, barbecues, water, showers; walking tracks and hiking trails; lookout; disabled access; interpretation centre.

Restrictions: No camping; no domestic animals.

track were created by the gradual cooling and contraction of lava. More can be seen at low tide near the shoreline.

The basalt soils which cover much of the park support a mix of dry and subtropical rainforest. Many rare and restricted plant species grow in the rainforests and a small area of heathland occurs on the exposed eastern slopes of the park.

Pandanus (*Pandanus tectorius*)

Brush-turkey
mound

On the western edge of the park near the Gold Coast Highway, the vegetation is dry eucalypt forest with mostly Brush Box (*Lophostemon confertus*). Forest Red Gums (*Eucalyptus tereticornis*) border the northern edge of the park. On the steep slopes close to the shoreline on the eastern edge of the park are a few stands of Pandanus (*Pandanus tectorius*).

This park is one of the best places in which to see Australian Brush-turkeys (*Alectura lathami*) and their nests which appear as large mounds throughout the park.

Koalas can still be occasionally sighted inside the park although the long-term prospects for their survival are not good given the potential for road kills on the Gold Coast Highway and small size of the habitat.

Unfortunately, populations of most small ground mammals have been devastated due to the park's isolation and relatively small size. Predation by cats and dogs has also reduced fauna numbers, although many birds can be seen looking for fruit in the rainforest canopy. Birds of prey such as the White-bellied Sea Eagle (*Haliaeetus leucogaster*) hunt around the edges of the headland.

Just off Echo Beach on the main walking track are the remains of a large shell midden. This is the most visible of the midden sites of the Kombumerri people who referred to this area as *Jellurgal.* — **Stephen Poole**

Look out for: Brahminy Kites (*Haliastur indus*) cruising above Tallebudgera Creek; large numbers of butterflies in the rainforest during summer.

Fig roots

GLASSHOUSE MOUNTAINS

The Glasshouse Mountains are among the best known landmarks of southern Queensland. Small national parks protect four of the mountains — Beerwah (245 ha), Tibrogargan (291 ha), Coonowrin (113 ha) and Ngungun (45 ha). The others — Coochin, Cooee, Tibberoowuccum, Tunbubudla, Beerburrum and Wild Horse Mountain — are contained within State Forestry Reserves.

The Glasshouses reach an average height of 400 m above sea level with the highest, Mt Beerwah, at 556 m. The mountains are a series of remnant volcanic plugs and physically would not have changed much over many thousands of years. They are mostly composed of the igneous rocks rhyolite and trachyte. Through a process known as columnar jointing, the rhyolite forms distinct columns and these are most visible at Coonowrin and Ngungun.

The land between the mountains has long been given over to agriculture and exotic pine plantations but within the reserves several endemic plant species have survived. These include the Ngungun May Bush (*Leptospermum leuhmannii*).

Above: Rhyolite columns, Coonowrin

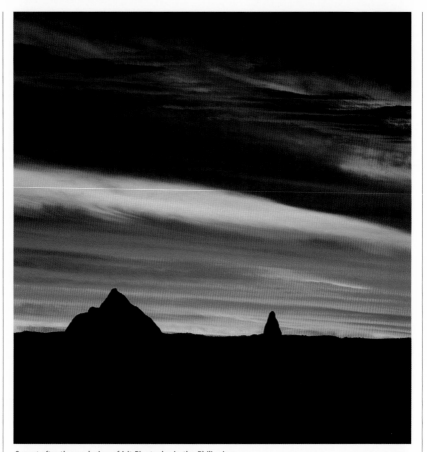

Sunset after the explosion of Mt Pinataubo in the Philippines.

The Glasshouses have always been popular with rock climbers although bushwalking is possible on most of the peaks, at least at lower elevations.

The mountains are spectacular when seen against the backdrop of the surrounding countryside in the early morning or late afternoon. Good viewing positions are the Interpretation Shelter at Wild Horse Mountain or along the escarpment road near Mary Cairncross Park at Maleny. — **Stephen Poole**

Look out for: Peregrine Falcons (*Falco peregrinus*) which nest high on the mountains.

MORETON ISLAND

Moreton Island is undeniably the jewel of the offshore islands. The largest of the Bay sand islands, Moreton is considered a near wilderness. This sense of "wilderness" comes from the island's size, its dense vegetation and the quality of its habitats. It is not diminished by the fact that thousands of people visit the island throughout the year. Even in "peak" holiday times it is possible to escape the crowds.

Moreton is not pristine but has felt the impact of man for many centuries. Discarded shell heaps or middens are reminders of the island's long Aboriginal history. In European times, some timber was taken to build settlements and the introduction last century of horses and goats created feral herds which damaged sensitive ecosystems. However these last two do not detract from the overall quality of the island's environment or from its importance to the region.

Moreton is 38 km long and has a maximum width of about 9 km. It covers 185 sq km and is part of the southern Queensland sand mass which include Fraser, the Cooloola coast, Bribie and North Stradbroke. Most of the island is national park but there are a few small settlements at Kooringal, Cowan Cowan and Bulwer. A resort at Tangalooma has a lease on the site of a former whaling station.

The island's topography is mostly rolling sand dunes and steep sand cliffs. Much of Moreton is covered by wallum heath, but there are also eucalypt forests, which in some areas are quite tall, swamps and sand dune communities.

One of the highest sand dunes in the world — Mt Tempest — is located in the centre of the island between

LOCALITY GUIDE

Location: At its closest point, Moreton is only 22 km from Brisbane GPO; 1 hour by launch or 2 hours by vehicular ferry.

Access: Ferries depart daily from Pinkenba, Lytton and Scarborough; boat anchorages are located on western side of island.

Facilities: Camping, water, toilets, showers; walking tracks and hiking trails; lookouts.

Restrictions: No domestic animals; camping and 4WD permits required; roads suitable for 4WDs only.

Near Five Hills

Cowan Cowan and Tangalooma, west of Eager's Swamp. Mt Tempest is 280 m high and the nearby Storm Mountain is 274 m. There are three dune blowouts (sand blows) on the island — the Desert, near Tangalooma, and the Big and Little Sandhills toward the southern tip. Sand blows are caused by disturbance to the vegetation cover and can become larger over time forming huge horse-shoe shapes.

Cape Moreton, the most north-easterly point, is the only rocky outcrop on the island and from here, land-based whale watching is possible during the northern and southern migration of the Humpback Whale (*Megaptera novaeangliae*) in April and October.

Mixed eucalypt forests and woodlands dominate the centre of the island across its entire width. Swamps and lakes occur in the extreme southern and also northern part of the island.

Bulwer Swamp is the largest of the freshwater swamps and is located near Comboyuro Point, the north-west corner of the island. Sedges and reeds are the most significant vegetation here but wildflowers like the Christmas Bell (*Blandfordia grandiflora*) also colour the landscape.

Moreton Island's closeness to Brisbane makes it particularly popular. The island maintains a fragile balance between the many plant communities on its sand-based environment. The Department of Environment recommends minimal impact practices — behaviour aimed at preserving ecological and wilderness values. Delicate dunes are protected by spinifex grass and ground covers — vital in preventing erosion and protecting nesting birds like Red-capped Plovers. Keep off dune grasses at all times. Lakes on the island are pure and contain few nutrients. Don't use sun-screens, soaps, shampoos or detergents in or near them. Use fireplaces where provided, not an open fire. Wildfires can be devastating. Remember good planning is the key to a successful visit with minimal impact.

Ophioscincus truncatus

Opposite: Big Sandhills **Above:** Sunrise, Ocean Beach

The largest freshwater lake on Moreton is Blue Lagoon, located on the eastern side of the island, a few kilometres south of Cape Moreton. Keep a lookout for insectivorous sundews (*Drosera binata*) around the edges of the lagoon (see p. 117).

Honeyeater Lake, another freshwater lake near Blue Lagoon is a noted cormorant roost and breeding area for the Musk Duck (*Biziura lobata*).[2]

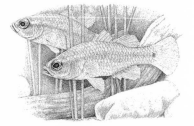

Oxleyan Perch *Nannoperca oxleyana*

Top: Blue Lagoon **Above:** Swamp, near Warrajamba Beach

Top: Warrajamba Beach
Above: Foxtail Ferns *Caustis blakei*

As the name suggests this is also a good location to observe honeyeaters. (For the adventurous, the less accessible Lake Jabiru area just south-west of Cape Moreton is a haven for water and wading birds.)

There are more than 500 species of plants on Moreton,[3] however most of the island is characterised by the banksia shrubs and trees of the wallum heath including Wallum Banksia (*Banskia aemula*) and Coast Banksia (*B. integrifolia*).

The low heathlands are excellent locations to see wildflowers, particularly during winter when White-cheeked and Brown Honeyeaters (*Phylidonyris nigra* and *Lichmera indistincta*) flock to the nectar-laden blossoms. As in similar vegetation communities on the mainland, stunted "mallee" eucalypts (see p. 23) also grow here. Pink Bloodwood (*Corymbia intermedia*),

Planchon's Stringybark (*E. planchoniana*), and Scribbly Gum (*E. racemosa*) all have mallee forms on Moreton.

The island's dry eucalypt forests include Blackbutt (*E. pilularis*), and Forest Red Gum (*E. tereticornis*). There are also large areas of Cypress Pine or Bribie Island Pine (*Callitris columellaris*) in the south of the island, near the Little Sandhill and fringing the Desert.

The rainforest tree Satinay (*Syncarpia hillii*) grows in sheltered gullies to the north of Mt Tempest[4] and Smooth-barked Apple (*Angophora leiocarpa*) and Smudgee (*A. woodsiana*) are common on the western side of the island. Fringing mangroves grow only

Above: Soldier crabs **Opposite:** North Point

on the western side of Moreton in a few unconnected pockets north of Reeders Point.

There are no large mammals such as kangaroos or wallabies on Moreton, but the complexity of the island's vegetation sustains a wide range of fauna from the unusual "acid frogs" (see p. 118) through to some of the most common animals of the Greater Brisbane Region.

The Tommy Round-head (*Diporiphora australis*) is an appealing lizard which reaches its southern limit of distribution in the Greater Brisbane Region and which is frequently encountered on Moreton. To the casual observer it looks like a smaller or juvenile version of the more common Bearded Dragon (*Pogona barbata*) but the Tommy Round-head is paler and lacks the row of spines along its back.

A chance encounter with a Sand Goanna (*Varanus gouldii*) on Moreton is an opportunity to note the colour and marking differences between this and the mainland goanna species, the Lace Monitor (*Varanus varius*).

Moreton Bay is home to three large oceanic turtles — the Logggerhead (*Caretta caretta*), Leatherback (*Dermochelys coriacea*) and Green (*Chelonia mydas*) — which can occasionally be seen on the beaches of the sand islands at night or in the offshore waters.

For diving enthusiasts, sunken wrecks — barges, dredges and sundry cargo vessels — have formed artificial reefs on the western side of Moreton at Cowan Cowan, Tangalooma, just north of the Big Sandhill and at Bulwer. It is ironic that two whaling boats sunk off Cowan Cowan now nurture nature. — **Stephen Poole**

Look out for: Beach-washed sea snakes; colourful dragonflies flitting over the lakes and swamps.

MT MEE STATE FOREST

Mt Mee State Forest is a pleasant one and half hours from the heart of Brisbane, far enough away for a day's outing in the countryside, yet close enough not to have to worry about extra provisions.

The forest is part of the northern D'Aguilar Range, in the Mt Mee complex which forms an elevated plateau system with steep escarpments and gorges to the west. The geology of Mt Mee is complex. The extensive state forest covers a number of geological formations and three major rock types — Bunya phyllites, Neranleigh-Fernvale beds and Rocksberg greenstone. The headwaters of the Stanley and North Pine Rivers and Neurum, Oaky and Reedy Creeks form in Mt Mee.

The forest is steeped in history. The Mt Mee area and parts of the D'Aguilar Range to the south are believed to have been inhabited by three Aboriginal groups — the Niablo, the Garumngar and the Dungidau — each of whom are thought to have spoken different dialects of the Wacca language.

Aboriginal words were used by European settlers to name various landmarks in the region. Although the local Aboriginal name for Mt Mee was *Dahmongah*, meaning flying squirrel, this was not assumed by settlers who chose to use Mt Mee as a derivation of *mia mia*, meaning view or lookout. Another local name, Neurum, also has Aboriginal roots. *Neurum*, meaning to sleep, is also believed to be the name of magic bone dust or crystal

Location: 60 km north-west of Brisbane GPO, 25 km from Dayboro; about 1.5 hours' driving.

Access: Via Samford, Dayboro and then Sellins Road before the Mt Mee township.

Facilities: Picnic areas, tables, barbecues, toilets, water; camping; walking tracks; trail bike riding; swimming; 4WD access; disabled access.

Restrictions: Permits are required for access beyond the camping and day use areas. Dogs must remain on a lead; dogs are not permitted in the camping area.

which was scattered on an enemy's bed to produce sores. These sores were called *neurum neurum*. It is assumed that the crystals were found in Neurum Creek or near there.

Early European timbergetters had a plethora of timber to choose from at Mt Mee. Initially, as was the case throughout Queensland, Red Cedar (*Toona ciliata*) was one of the first timbers logged. As this resource dwindled, other species were selected — Hoop Pine (*Araucaria cunninghamii*), and various eucalypts.

Above: Emerald-spotted Treefrog *Litoria peronii* **Opposite:** Piccabeen Palms (*Archontophoenix cunninghamiana*)

For a time the timber industry boomed at Mt Mee with most being taken to Brisbane for housing and construction. Bullock teams were used to haul the logs to sawmills and consequently many local features — such as Tommy's Knob and Top Yard Road — were named for or by the bullockies who worked the region.

Timber from Mt Mee was used in Mackay's harbour wharves, on the Hornibrook Bridge at Redcliffe and in St Stephen's Cathedral in Brisbane. The area still provides timber to Brisbane markets.

In the period following World War I, the government cleared areas of the state forest for settlement. In 1934, 32 families settled in Centipede Creek Valley to grow bananas. However, after initial success this venture failed and the families began to leave. The land was resumed back into state forest in the 1970s.

Rocky Hole

Mt Mee has much to offer visitors. The Gantry Day Use Area is an attractive picnic spot surrounded by wet eucalypt forest. It is named for its huge gantry structure — the remains of the local sawmill which operated until 1981.

Western escarpment, view to Brisbane Forest Park

Scribbly Gum forest

Wet eucalypt forests usually have tall trees with some rainforest species in the understorey. These forests occur where annual rainfall averages between 1500 mm and 2000 mm and the soil is moderately fertile (see p. 21). Blackbutt (*Eucalyptus pilularis*) dominates the wet eucalypt forests in this area and other species include Flooded Gums (*E. grandis*) and Red Bloodwood (*Corymbia gummifera*).

Fauna around the picnic grounds is fast and shy but patient and quiet watchers may be able to capture a glimpse of a Land Mullet (*Egernia major*), Burton's Snake Lizard (*Lialis burtonis*), Golden Whistler (*Pachycephala pectoralis*), White-throated Treecreeper (*Cormobates leucophaea*) or White-browed Scrubwren (*Sericornis frontalis*). From the Gantry Day Use Area, a short drive along the Neurum Creek Road enables visitors to appreciate some of the forest's varied habitats and vegetation.

A small Hoop Pine plantation and grazing cattle demonstrate the principles of multiple land use management as applied to some state forests. This effectively means that the same area is used for a variety of purposes considered compatible with each other.

In the Bull Falls area, there is a small patch of regenerating dry rainforest growing on shallow, basaltic soil.

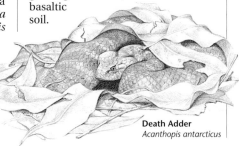

Death Adder
Acanthopis antarcticus

Dominant trees are Red Ash (*Alphitonia excelsa*) and Crow's Apple (*Owenia venosa*) with pioneer species such as Mulberry Leaf Stingers (*Dendronicide photinophylla*), Wild Tobacco (*Solanum mauritianum*) and Castor Oil Plants (*Ricinus communis*).

Vines and shrubs proliferate in the rainforest. Native Indigo (*Indigofera siliata*), Dogwood (*Jacksonia scoparia*), Native Holly (*Alchornea ilicifolia*) and Currant Bush (*Carissa ovata*) form a seemingly dense barrier in the understorey.

The view from the lookout platform at Bull Falls is of rainforest scrub. If you look carefully you may be able to see vines among the canopy trees, King Orchids (*Dendrobium speciosum*), emerging Hoop and Bunya Pines (*Araucaria cunninghamii, A. bidwilli*) and maybe even the vibrant red flowering of Flame Trees (*Brachychiton acerifolius*). The falls do not flow during dry weather but are spectacular after heavy rain.

Red-necked Pademelons (*Thylogale thetis*), Eastern Whipbirds (*Psophodes olivaceus*), Brown Cuckoo-doves (*Macropygia amboinensis*), Yellow-faced Whip Snakes (*Demansia psammophis*) and Wood Cockroaches (*Panesthia cribrata*) are some of the rainforest animals that visitors might encounter.

Further along the Neurum Creek Road the sub-tropical rainforest begins to merge with wet eucalypt forest again dominated by Flooded Gums (*Eucalyptus grandis*).

The Mill Rainforest Walk is a 1.3 km graded walk which passes through sub-tropical rainforest and the dynamics of rainforests are explained in signage along the walk. One of the most recognisable sounds of the rainforest is the unique call of the Green Catbird (*Ailuroedus crassirostris*) — rather like a baby crying.

Some 600 ha of Hoop Pine plantations are included in the Mt Mee State Forest. Hoop Pine is one of Australia's most impressive native conifers but pure stands are rarely found. As seedlings in the rainforest Hoop Pine remain suppressed in shaded situations. It is not until a break in the canopy occurs releasing sunlight that the seedling is able to become a forest giant growing to 60 m. Hoop Pines in plantations are grown for up to 50 years before harvest. The tree takes its common name from the way in which the bark curls into hoops as it is shed from the tree.

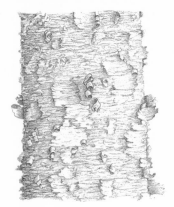

Hoop Pine *Araucaria cunninghamii*

Above: Early morning mist

The Neurum Camping area, about 2 km on from the Mill Rainforest, is set among the dry eucalypt forest which covers much of Mt Mee. This vegetation community is characterised by tall eucalypt species with an understorey of grass or shrubs (see p. 19). Species include Grey Ironbark (*E. siderophloia*), Stringybark (*E. eugenioides*), White Mahogany (*E. acmenoides*), Tallowwood (*E. microcorys*) and Spotted Gum (*Corymbia variegata*).

Some of Mt Mee's unique forest types such as the Scribbly Gum (*E. racemosa*) communities on the western escarpment are in restricted access areas so drivers need a Permit to Traverse to venture further afield. — **Shawn Delaney**

Look out for: Crimson Rosellas (*Platycercus elegans*) flashing over the dark green forest canopy; vivid blue dragonflies hawking insects over rock pools.

Flooded Gum

NERANG STATE FOREST

Many millions of years ago, movement of the Earth's crustal plates near the edge of what was, at the time, the eastern coast of Australia resulted in compression of deep water sediments. The subsequent upward thrusting of these sediments formed the mountainous terrain of which Nerang State Forest is part.

The reserve covers about 1750 ha and includes the headwaters of Coombabah Creek. Dry eucalypt forest is the dominant vegetation and the forests typically contain Grey Gum (*E. biturbinata*), Narrow-leaved Ironbark (*E. crebra*), Spotted Gum (*Corymbia variegata*) and Red Bloodwood (*C. gummifera*).

There is also a small section of dry rainforest which is an important remnant of a previously greater distribution of this forest type. The dry rainforest can be easily identified by the Hoop Pine (*Araucaria cunninghamii*) which protrudes above the canopy. The relatively rare *Cassia marksiana* can be found within the dry rainforest, along with another species, *Microcitrus australasica*, which is at its northern limit here.

Several sightings have been made of a Powerful Owl (*Ninox strenua*) in the forest. The owl preys on birds and mammals — its diet consisting mainly of small and medium-sized mammals, especially the Greater Glider (*Petauroides volans*) and the Common Ringtail Possum (*Pseudocheirus peregrinus*).

A visitor to Nerang may also be fortunate enough to see one or more of the four wallaby species which have been recorded in the forest and surrounding areas. These include the Red-necked Wallaby (*Macropus rufo-*

Inset opposite: Common Ringtail Possum
Pseudocheirus peregrinus

griseus), the Agile Wallaby (*M. agilis*), Swamp Wallaby (*Wallabia bicolor*) and the Golden Swamp Wallaby (*W. bicolor welsbyi*).

In addition to catering for the more traditional forestry uses such as wood and honey production, Nerang State Forest is managed as an active nature-based recreation area where activities such as mountain bike riding, horse riding and bushwalking are encouraged. These are all great ways to experience a rich and diverse landscape. — **Mark Peacock**

Look out for: Sleepy koalas (*Phascolarctos cinereus*) napping high in the branches of eucalypt trees.

NORTH STRADBROKE ISLAND

Many Brisbane residents escape to North Stradbroke Island to experience a little "wilderness" but with some facilities and comforts. The island, which is about 32 km long and 11 km wide, is known for its outstanding beauty and because of this, much of it is listed on the register of the National Estate.

Despite extensive silica mining, the island still has dune vegetation, swamps, rainforests, eucalypt forests, mangroves and saltmarshes, and perched lakes.

Apart from some rocky outcrops at Point Lookout (rhyolite) and Dunwich (sandstone), sand dune systems dominate the island. The spectacular banksia blooms of the heath vegetation on the eastern side of Stradbroke are a "must see". Saw or Old Man Banksia (*Banksia serrata*) is a common heathland plant. This species grows up to 12 m tall and has large yellow flowers up to 14 cm long. The leaves have a distinctive saw-tooth serration.

Stradbroke is also known for its peat swamps or peat bogs as they are more commonly called. Eighteen Mile Swamp on the eastern side of the island is the largest of its type in South-east Queensland. The peat layer in these swamps is about a metre deep and the water table is often close to the surface.

Water Mouse
Xeromys myoides

LOCALITY GUIDE

Location: About 40 km from Brisbane GPO.

Access: Car ferry departs Redland Bay to Dunwich; or water taxis depart from Cleveland to Dunwich. Rail travel to Cleveland and water taxi can mean a travel time of only an hour and a half from Brisbane.

Facilities: All facilities outside national park.

Restrictions: No domestic animals.

Opposite: Brown Lake **Above:** Cabbage Tree Palms (*Livistona australis*), Eighteen Mile Swamp

A walk through the peat swamps can be a little like struggling through mudflats. The beautiful Cabbage Tree Palm (*Livistona australis*) grows along the southern edge of the Eighteen Mile Swamp.

The remnant rainforest at Myora is one of the gems of North Stradbroke and many species of plants and animals — for example, the Purple-crowned Pigeon (*Ptilinopus superbus*) — are believed to be restricted to this one location.

The eucalypt forests and woodlands on the island are dominated by Blackbutt (*Eucalyptus pilularis*), Scribbly Gum (*E. racemosa*), Queensland White Stringybark (*E. tindaliae*) and Red Mahogany (*E. resinigea*) and the related Pink Bloodwood (*Corymbia intermedia*). Swamp Mahogany (*E. robusta*) grows in the peat swamps and Planchon's Stringybark (*E. planchoniana*), a restricted species around Brisbane, is also found on the island.

All seven mangrove species found in the Greater Brisbane Region occur on North Stradbroke — the Grey (*Avicennia marina*), Red (*Rhizophora stylosa*), Orange (*Bruguiera gymnorrhiza*), River (*Aegiceras corniculatum*), Milky (*Ex-*

Top: Cylinder Headland **Above:** Myora, sandflats to freshwater swamp

coecaria agallocha), Black (*Lumnitzera racemosa*) and the Spurred (*Ceniops tagul*). The Grey Mangrove is the most common and the mangroves are concentrated on the western shore of the island.

As would be expected of a sand island, there are numerous freshwater lakes — the best known being the Brown and Blue Lakes. Blue Lake is the only national park on the island and is relatively small, about 50 ha.

Sometimes islands form ecological refuges where species that have been lost elsewhere are able to survive. This is because islands often do not suffer the same environmental disturbance and degradation as other places and the numbers of predatory animals may be lower than in urban areas. North Stradbroke is one of these "ecological islands" and supports a number of restricted and rare plants and animals.

The Water Daisy (*Olearia hygrophila*) is endemic to the island and not known from any other location.[5] Christmas Bells (*Blandfordia grandiflora*) occur in only a few places in South-east Queensland including North Stradbroke. Pandanus or Screw Palms (*Pandanus tectorius*) grow around Point Lookout. These plants are strand vegetation and are found on only a few headlands in the Greater Brisbane Region.

North Stradbroke has the most diverse fauna of the Bay islands and this suggests that the island was connected to the mainland more recently (in geological time) than Moreton.

It is on the outer edge of the peat swamps on the western side of the island, north of Dunwich and the rainforest at Myora that the Water Mouse (*Xeromys myoides*), one of Australia's rarest animals has been found — from the swamp line through the saltmarsh to the mangrove communities at low tide.

Myora

North Stradbroke supports three species of carnivorous Sundews (*Drosera binata, D. peltata* and *D. spatulata*). These "insect-eating" plants secrete a sticky substance in which insects become trapped, to be later "digested" by the plant. However, a bug of the family Miridae is able to move across the plant's surface safely and browse on the plant and trapped insects.[7] — SP

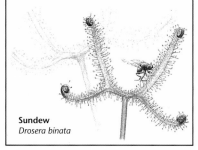

Sundew
Drosera binata

Above: Brown Lake **Opposite:** Aerial view, Eighteen Mile Swamp **Insets:** Paperbark and banksia blossoms

The only population of the Agile Wallaby (*Macropus agilis*) found south of Rockhampton survives on North Stradbroke and Peel Islands. The Bush hen (*Amaurornis olivacea*) is at its southernmost limit on North Stradbroke.

Sometimes the isolation of islands can also lead to the development of variations or new forms within species. On North Stradbroke, one of the most noticeable examples of this process is the golden colour form of the Swamp Wallaby (*Wallabia bicolor welsbyi*). The Swamp Orchid (*Phaius australis* var. *benaysii*), which is the largest ground orchid in Australia with flower spikes up to 2 m long, has a "yellow" colour form on Stradbroke.

Koalas can be found across the island, though numbers are low. Sixteen species of snakes have been recorded on the island. Of the dangerous snakes, the most common is probably the Death Adder (*Acanthopis antarcticus*). — **Stephen Poole**

Look out for: Whales and sea birds offshore from Point Lookout; Fire-tailed Skinks (*Morethia taeniopleura*) in the leaf litter around Wallum Banksias.

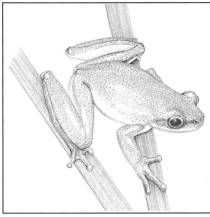

A number of frogs occur on North Stradbroke and Moreton Islands which are known collectively as the "acid frogs". These frogs have adapted to the acidic wetlands of the sandy coastal lowlands ("wallum") of southern Queensland and northern New South Wales. One of these frogs, the Cooloola Sedgefrog (*Litoria cooloolensis*) reaches its southernmost distribution on North Stradbroke and is known only from the Brown Lake area and the swamps on Moreton Island. — SP

Cooloola Sedgefrog

NUMINBAH STATE FOREST

The topography of Numinbah State Forest is the result of a turbulent geological history which included periods of large scale sinking of the landscape and subsequent violent volcanic eruptions. The volcanic plugs formed as a results of this activity include Egg Rock and Page's Pinnacle and they remain striking physical elements in the Numinbah Valley.

The rugged terrain characteristic of Numinbah State Forest supports a rich flora and fauna. The steeper slopes and ridges contain vegetation communities dominated by eucalypts. Common species are Grey Gum (*Eucalyptus biturbinata*), White Mahogany (*E. acmenoides*) and Red Bloodwood (*Corymbia gummifera*).

LOCALITY GUIDE

Location: 85 km south of Brisbane GPO; 1 hours' driving.

Access: Follow the Nerang-Murwillumbah Road from Nerang to the state forest.

Facilities: Picnic area, tables, shelter shed, toilets, water, barbecues; disabled access; swimming; track network for walkers, horses and bikes.

Restrictions: Dogs must remain on a lead; restricted vehicle access.

Nerang River

In some of the wetter and more protected areas, particularly along parts of the Nerang River and Waterfall Creek, patches of tall "closed" forest, where the trees grow close together, can be discovered. These wet eucalypt forests may contain trees up to 50 m tall such as Flooded Gum (*E. grandis*), Black Bean (*Castanospermum australe*) and Silky Oak (*Grevillea robusta*).

Australia's largest bird of prey, the Wedge-tailed Eagle (*Aquila audax*) is known to nest in Numinbah State Forest. The number of these birds across Australia is believed to be relatively low and this is probably the direct result of the bounties which were paid to destroy them in the past.

Australia's most recognisable monotreme mammal, the Platypus (*Ornithorhynchus anatinus*) has also been recorded at Numinbah. This shy creature is unlikely to be seen by most visitors, however a lucky few may see the animal meticulously grooming itself on a rock or a log or foraging for food below the surface of waterways

The picturesque Nerang River winds its way through the forest and is popular for swimming or just relaxing beside its clear waters. Numinbah State Forest is part of the Hinze Dam catchment area which supplies water to Gold Coast City residents. As well as to its catchment values, the area is managed to provide recreation, wood and honey production, conservation and education benefits.

For the more adventurous visitors, great views and spectacular forests can be discovered by walking, riding a horse or a even a mountain bike through some of the more remote parts of the forest. — **Mark Peacock**

Look out for: A glimpse of the Firetail Gudgeon (*Hypseleotris galii*) or the Eel-tail Catfish (*Tandanus tandanus*) in the Nerang River or Waterfall Creek.

Top: Forest canopy **Centre:** Black Bean seed pods **Below:** Casuarinas

TAMBORINE MOUNTAIN

Tamborine Mountain is a plateau more than 500 m above sea level which runs north from the McPherson Ranges on the Queensland/New South Wales border. It is part of the Darlington Range and was the northern lava flow of the Tweed Shield Volcano (see p. 95).

The Tamborine Plateau has been densely settled over many years and this has greatly affected its natural habitats. However a few, scattered areas have been incorporated into the Tamborine National Park. Because most of the walking trails in the parks are short, Tamborine is ideal for seeing a wide variety of vegetation and fauna over relatively small distances.

The national park comprises 17 sections, the most significant being Witches Falls, Palm Grove, MacDonald, Lepidozamia Grove, Cedar Creek, The Knoll and Joalah and Panorama Point. All are relatively small (Panorama Point is undeveloped.) Although heavily used by people escaping the summer heat in Brisbane and the Gold Coast, Tamborine National Park still harbours some wonderful natural gems.

It protects sub-tropical and dry rainforest, dry eucalypt forest and easily accessible wet eucalypt forest. There is rainforest in nearly all sections of the park. Walkers in the sub-tropical rainforest will come upon enormous examples of Strangler Fig (*Ficus watkinsiana*) and the giant buttressed Yellow Carabeen (*Sloanea woollsii*), particularly at Palm Grove, Witches Falls and Joalah. (Had Queensland's first national park, Witches Falls, not been gazetted in 1908, many of these trees may not have survived.) Epiphytic plants such as staghorn and elkhorn ferns and many orchids grow on the rainforest trees and in the wet eucalypt forests.

LOCALITY GUIDE

Location: 80 km south of Brisbane GPO; 1 hours' driving.

Access: 10 km west of Pacific Highway, via Beenleigh and Tamborine Village or Oxenford.

Facilities: Picnic grounds at most parks, tables, shelter sheds, toilets; walking tracks; lookouts; interpretation centre; disabled access.

Restrictions: No domestic animals; no camping,

Epiphytes are plants which live in a symbiotic relationship on other plants (i.e. both species gain from the relationship and the host tree is not killed). Orchids, Bird's Nest, Elkhorn and Staghorn Ferns and the like are all epiphytic as are lichens and mosses. In sub-tropical rainforest and wet eucalypt forest, the epiphytes growing high in the canopy differ from those growing on the lower trunks and branches. Epiphytes can become so large and heavy that from time to time, they fall to the ground. — SP

King Orchid *Dendrobium speciosum*

Above: Wet eucalypt forest   **Below:** Cedar Creek

A massive Moreton Bay Fig (*F. macrophylla*) can be found not far from the start of the Palm Grove walking trail. Tamborine has a number of stands of Piccabeen Palms (*Archontophoenix cunninghamiana*) — most obviously at Palm Grove, but also at Witches Falls, MacDonald, The Knoll and Joalah.

As its name suggests, Lepidozamia Grove is noted for the ancient cycad *Lepidozamia peroffskyana*, a species which is believed to have existed unchanged for up to 300 million years.

Within the dry eucalypt forest that occurs across the top of the plateau, particularly near cliff edges, the Sydney Blue Gum (*Eucalyptus saligna*) is a notable species. Small areas of dry rainforest also grow at the base of the cliffs along Cedar Creek.

Tamborine National Park also protects outstanding areas of wet eucalypt forest dominated by Flooded Gum (*E. grandis*) with palm groves forming the understorey.

Tamborine's wildlife is as varied as the vegetation. One of the most unusual inhabitants of the mountain is the giant greyish earthworm *Digaster longmani*. These worms can grow to more than a metre in length and up to 3 cm in diameter. The worm makes a bubbling sound as it moves through rain-saturated soil. [6]

Tamborine wildlife is often easily seen — birds in particular frequent the edges of the forests and scavenge the picnic areas. The distinctive calls of the black and white Pied Currawong (*Strepera graculina*) and Eastern Whipbird (*Psophodes olivaceus*) can be heard across Tamborine. The crested Whipbird is mostly black and is sometimes encountered close to or on walking tracks. — **Stephen Poole**

Look out for: Brightly coloured rainforest fruits on the ground; pademelons grazing on the edges of forests.

Tamborine National Park is one of the most studied national parks in Queensland due to the efforts of the voluntary Tamborine Natural History Association. Visit their information centre at Doughty Park. The association has a number of fauna specimens on display along with seed pods and fruit of various trees. It also produces leaflets on the flora and fauna of Tamborine, especially birds.

Above: Sub-tropical rainforest **Centre:** Spotted Gum *Corymbia variegata* **Below:** Near Witches Falls

WHITE ROCK

White Rock is the major feature of the recently declared White Rock Conservation Park. While the area is relatively close to the cities of Brisbane and Ipswich it is somewhat unknown and unexplored by the general community.

Because the conservation park is relatively new White Rock has no facilities or signage to assist visitors negotiate the numerous tracks and people have become disorientated while seeking White Rock. It is suggested that visitors be accompanied by someone who knows the area.

It takes about 50 minutes to walk to White Rock through a number of habitat types. These include a large stand of paperbarks, tall eucalypt forests with a mixture of understoreys, sandy flats, and a ferny gully bounded by steep rocky escarpments. On the high ridges there are examples of heathland vegetation characterised by stunted trees growing out of crevices in solid rock.

White Rock has some spectacular large rocky outcrops, sandstone ridges and boulders in which lie numerous wind-formed caves. The rock and other vantage points offer remarkably wide views of the adjoining Woogaroo Creek and Six Mile Creek catchments, as well as Brisbane City, Moreton Bay, Stradbroke Island and several mountains.

LOCALITY GUIDE

Location: 30 km south-west of Brisbane GPO; 40 mins' driving.

Access: 4 km south of Redbank Plains, via Redbank Plains School Road to carpark at the entrance to Six Mile Creek Conservation Park.

Facilities: None.

Restrictions: No domestic animals; no 4WD access.

The rich sandstone features of the area have an attraction which often belies the harshness of this rugged country. But for the unprepared, the steepness of the ridgelines, the sparse trees and lack of cool respite and facilities can be oppressive in the summer months. Winter and spring are a different experience and offer a bounty of wild flowers and wildlife.

Woodland birds abound in the area and few local spots offer the species diversity known from the various habitats. These combinations are what make White Rock and its surrounds so unique and its flora and fauna so appealing to anyone in pursuit of an adventure in a "wild place".

Black-striped Wallaby
Macropus dorsalis

Above: Sandstone formation **Opposite:** Bushland to the south

White Rock also has a significant Aboriginal history and a "feel of ancient times" (see p. 131). While there has been debate about how sacred or significant the area is to Aboriginal people, there is no doubt that the area has a rich cultural history. It is unfortunate that the local Aboriginal tribes were dispersed and much of their history with them when European settlers arrived in the region. If visiting White Rock, respect the area's historical importance and treat it accordingly.

During World War II Australian and American soldiers were trained in the region and White Rock was used by the Americans as a command post to conduct and oversee training operations because of its advantageous views. Evidence of these activities can still be found throughout the area in the form of sandstone gunpits and used and unused ammunition cartridges.

The vegetation of the White Rock area is well documented and some unique species have been located. Plunkett Mallee (*Eucalyptus curtisii*) is a small tree widely used in landscaping for its beautiful and abundant creamy flowers. In the wild, it is known to exist in only a few scattered locations in South-east Queensland. *Marsdenia coronata* is an impressive vine which was once presumed extinct and is

Above: White Rock

listed as endangered and at risk of disappearing from the wild within 10 to 20 years. White Rock is possibly the best chance this species has of remaining known as a wild plant.

Tiny Greenhood (*Pterostylis* spp.), a diminutive terrestrial orchid, is also known only from a few sites. Bottlebrush Grass Tree (*Xanthorrhoea macronema*) and pink-flowering Native Hibiscus (*Hibiscus heterophyllus, H. splendens*) are usually found growing in coastal districts but at White Rock both can be seen growing along the sandstone outcrops.

White Rock's fauna is as impressive as its vegetation, but is often elusive. Depending on the season there are numerous birds. The Peregrine Falcon (*Falco peregrinus*), Grey Goshawk (*Accipiter novaehollandiae*), Powerful Owl (*Ninox strenua*), Glossy Black Cockatoo (*Calyptorhynchus lathami*), Black-chinned Honeyeater (*Melithreptus gularis*) and Spotted Quail-Thrush (*Cinclosoma punctatum*) are all known in the area.

Among the many mammals that it is possible to encounter are the Black-striped Wallaby (*Macropus dorsalis*) and Brush-tailed Rock Wallaby (*Petrogale penicillata*). Reptiles are abundant and two of the more common

species are the Spotted Velvet Gecko (*Oedura tryoni*) and Burton's Snake Lizard (*Lialis burtonis*). Visit the area on a wet summer's night and it is possible to hear, and perhaps see, some 20 species of frog. — **Adrian Caneris**

Look out for: Echidnas and koalas; dingoes, especially early in the morning near waterholes.

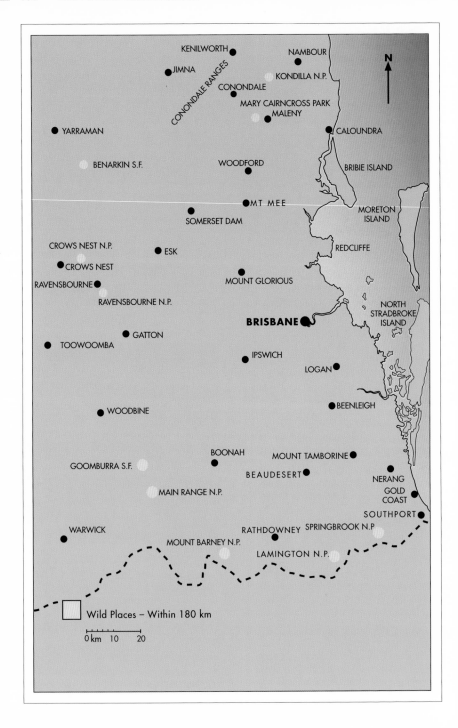

KENILWORTH

NAMBOUR

JIMNA

KONDILLA N.P.

CONONDALE

CONONDALE RANGES

MARY CAIRNCROSS PARK

MALENY

YARRAMAN

CALOUNDRA

BENARKIN S.F.

WOODFORD

BRIBIE ISLAND

MT MEE

SOMERSET DAM

MORETON ISLAND

CROWS NEST N.P.

ESK

REDCLIFFE

CROWS NEST

RAVENSBOURNE

MOUNT GLORIOUS

RAVENSBOURNE N.P.

NORTH STRADBROKE ISLAND

BRISBANE

GATTON

TOOWOOMBA

IPSWICH

LOGAN

WOODBINE

BEENLEIGH

BOONAH

MOUNT TAMBORINE

GOOMBURRA S.F.

BEAUDESERT

NERANG

MAIN RANGE N.P.

GOLD COAST

SOUTHPORT

WARWICK

RATHDOWNEY

SPRINGBROOK N.P.

MOUNT BARNEY N.P.

LAMINGTON N.P.

Wild Places – Within 180 km

0 km 10   20

N

WILD PLACES —
WITHIN 180 KM

Sunrise, Mt Barney

BENARKIN STATE FOREST

Benarkin State Forest covers some 16,000 ha along the scenic Blackbutt Range and includes sub-tropical rainforest, eucalypt forest and pine plantations.

Hoop Pine (*Araucaria cunninghamii*) occurs naturally in the area and is a common emergent species (protrudes above the canopy) of sub-tropical rainforests. Benarkin is so well suited to Hoop Pine that during the 1920s the Forestry Department established extensive plantations which now form an important timber resource for local and regional communities.

Visitors to Benarkin State Forest will notice that it has a strong forestry production emphasis. Hoop Pine plantations dominate much of the landscape, however the forest has more to offer. The eucalypt forests are characterised by Blackbutt (*Eucalyptus pilularis*), Tallowwood (*E. microcorys*), Grey Gum (*E. biturbinata*) and White Mahogany (*E. acmenoides*).

LOCALITY GUIDE

Location: 150 km north-west of Brisbane GPO: 2 hours' driving.

Access: Turn off the D'Aguilar Highway at Emu Creek Road adjacent to the Benarkin Day Use Area.

Facilities: Picnic areas, tables, toilets, barbecues, water, shelter sheds; two self-registration camping areas, (camping with dogs permitted in Clancy's Camp); walking tracks and hiking trails; lookouts; swimming; canoeing, forest drive; access to Bicentennial National Trail; disabled access; caravan access.

Restrictions: Dogs must remain on a lead; restricted vehicle access beyond camping area.

Opposite: Hoop Pines

Benarkin Forest is a haven for one of the largest and rarest of the button quails — the Black-breasted Button Quail (*Turnix melanogaster*). This sedentary bird is a clumsy flier and if disturbed, is more likely to freeze on the spot or run away than take to the air.

Yellow-bellied Gliders (*Petaurus australis*) have also been recorded in Benarkin. This species is one of Australia's most proficient gliders and they are considered to be particularly active and gregarious.

Picturesque Emu Creek meanders along the southern boundary of the forest and is ideal for swimming, canoeing or just relaxing. Both camping areas are situated on the banks of the creek. — **Mark Peacock**

Look out for: Grass Trees (*Xanthorrhoea* spp.) throughout the drier eucalypt forests; White-eared Monarchs (*Monarcha leucotis*) foraging in dry rainforest margins.

Above: Emu Creek

CONONDALE RANGES

The beautiful Conondale Ranges — an area remarkable for its wildlife, rugged mountain scenery, streams and waterfalls — are an easy two-hour drive from Brisbane.

The largest tract of sub-tropical rainforest and associated eucalypt forest remaining on Queensland's Sunshine Coast is found in the Conondales. The extensive native forests and about 4800 ha of Hoop Pine (*Araucaria cunninghamii*) plantation are encompassed by two national parks (7008 ha) and two state forests (53,000 ha).

Bunya Pines (*A. bidwilli*), Hoop Pines and large fig trees draped with epiphytic ferns and orchids can be seen soaring above the rainforest canopy. Piccabeen Palms (*Archontophoenix cunninghamiana*), diminutive Walking Stick Palms (*Linospadix monostachya*), tree ferns and the progeny of the canopy trees cover the forest floor.

Flooded Gums (*Eucalyptus grandis*) are a feature of watercourses and wetter parts, but it is in the drier, more fire-prone areas that the eucalypts dominate. These "drier" communities are prevalent in the north and west of the range, mainly in Jimna State Forest.

Left: Emergent Bunya Pines

Several species of uncommon or rare plants have been found here, including *Austromyrtus inophloea, Macadamia ternifolia, Acomis acoma, Sophora fraseri* and *Marsdeina coronata.* Other significant plants include the pipe vine *Aristolochia praevenosa* (the host plant for the endangered Richmond Birdwing Butterfly *(Ornithoptera richmondia)* and Forest She-Oak *(Allocasuarina torulosa)* the seeds of which form the diet of the Glossy Black Cockatoo *(Calyptorhynchus lathami).*

The fauna of the Conondale Range is similarly rich and diverse. Almost 30 different mammals and some 70 reptile species have been recorded here, but the area is most noted among naturalists for its frogs and birds (more than 100 species).

Fruit-doves, pigeons, parrots, honeyeaters and birds of prey are commonly seen throughout the Conondales. A number of rare and threatened species such as the Eastern Bristlebird *(Dasyornis brachypterus)*, Red Goshawk *(Erythrotriorchis radiatus)* and Glossy Black Cockatoo make their homes here. The wet eucalypt forests of the Conondales are considered to be critical habitat for the Marbled Frogmouth *(Podogarus ocellatus).*

The Conondales are also renowned as the home of two species of "disappearing" frogs. These are the Southern Platypus-frog *(Rheobatrachus silus)* (see p. 146) and Southern Dayfrog *(Taudactylus diurnus)* which have not been seen in the wild since about 1979 and are now feared to be extinct. Other uncommon or rare frog species include the Green-thighed Frog *(Litoria brevipalmata)*, Australian Marsupial Frog *(Assa darlingtoni)* and Fleay's Barred-frog *(Mixophyes fleayi).*

Platypus *(Ornithorynchus anatinus)* and Water Rats *(Hydromys chrysogaster)* frequent many of the larger watercourses. A quiet walk around dawn or dusk is the best chance to spot a platypus.

Top and opposite: Piccabeen Palms *(Archontophoenix cunninghamiana)*

Gliders, including the threatened Yellow-bellied Glider (*Petaurus australis*), occur throughout the Conondales. Mountain Brushtail Possums (*Trichosurus caninus*), Common Brushtail Possums (*Trichosurus vulpecula*), Bush Rats (*Rattus fuscipes*) and Fawn-footed Melomys (*Melomys*

cervinipes) are the mammals most likely to be seen around campsites. It is only at Jimna that kangaroos and wallabies become obvious.

Reptiles, fish and a host of invertebrates are also part of the Conondale Range fauna. Many of these animals are secretive and rarely seen but include the Elf Skink (*Eroticoscincus graciloides*), a spiny crayfish (*Euastacus hystricosus*), the Giant Earwig (*Titanolabia colossus*), Richmond Birdwing Butterfly and a large predatory ground beetle (*Castlenaudia porphyriacus*).

Although the national parks are undeveloped, camping and picnic facilities are available in the adjacent state forests of Kenilworth and Jimna.
— **Greg Czechura and Toni Hess**

KENILWORTH STATE FOREST

Booloumba Creek and Charlie Moreland camping and day use areas are an ideal base for exploring the Conondales.

From these two sites there are a range of forest walks for bushwalkers to enjoy. Short easy strolls such as the Fig Tree Walk from the Little Yabba rest area will lead to spectacular examples of Moreton Bay Fig Trees (*Ficus macrophylla*). Easy access makes these trees a "must see" on any visit to the region. The Mt Allan Hiking Trail is more challenging and has breathtaking views of Bellambi Gorge and surrounding forest.

LOCALITY GUIDE

Location: 175 km north-west of Brisbane GPO; 2 hours' driving.

Access: Take the Bruce Highway – Glasshouse Mountains scenic route — then travel via Landsborough through Maleny to the Sunday Creek and Booloumba Creek Roads.

Facilities: Picnic areas, toilets, tables, water, barbecues; walking tracks; self-registration camping, showers; forest drive; swimming; disabled access; caravan access.

Restrictions: Dogs are not permitted in the camping or day use areas.

A leisurely 37 km forest drive takes visitors into the heart of the Conondales. Pack a picnic and stop at Peter's Creek for lunch. Make the short walk to the rock pools set among towering Flooded Gums (*Eucalyptus grandis*). Other features along the drive include scenic lookouts and Bellambi Gorge which begins at the junction of Bellambi and Peter's Creeks. The creeks meet in a ruggedly attractive setting of cascades, falls and rock pools. The large rock outcrop between the creeks is called "The Bread Knife".
— Toni Hess

Opposite: Cascades, junction of Bellambi and Peter's Creeks **Above:** Bellambi Gorge

Look out for: Brush-turkeys (*Alectura lathami*) scratching in the undergrowth and Platypus (*Ornithorhynchus anatinus*) feeding in the creeks around dawn and dusk.

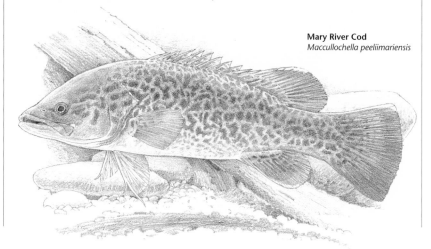

Mary River Cod
Maccullochella peeliimariensis

JIMNA STATE FOREST

Further along the Sunday Creek Road on the western side of the Conondale Ranges is Jimna State Forest. Jimna boasts the tallest fire tower in the state. The platform is some 44 m and 241 steps high and provides a 360 degree panorama, ideal for viewing sunrises and sunsets.

Peach Trees, on the grassy banks of Yabba Creek, is an excellent camping spot or base for exploring other points of interest around Jimna.

Walking tracks around Peach Trees pass through open forest, and dry rainforest, while other walks follow Yabba Creek. The Araucaria Walk leads to a variety of habitats including a rainforest characterised by ancient Hoop and Bunya Pines (*Araucaria cunninghamii, A. bidwilli*).

LOCALITY GUIDE

Location: 160 km north-west of Brisbane GPO; 2 hours' driving.

Access: Take the Caboolture exit off the Bruce Highway, follow the D'Aguilar Highway to Kilcoy and onto the Kilcoy-Murgon Roads to Jimna.

Facilities: Picnic area, toilets, barbecues, tables, water; self-registration camping, showers; walking tracks; lookouts; swimming; disabled access.

Restrictions: No dogs in camping or picnic areas; dogs on a lead are permitted only at Jimna Fire Tower.

Opposite: Dry eucalypt forest **Above:** Catchment, Sheep Station Creek

The Hoop Pine plantations are also worth exploring. The Araucaria Walk takes in a particular forest community known as "Bastard Scrub". (This is a poorly developed forest type which has rainforest species as a dense under-storey but lacks the conditions to develop any further.) While walking enjoy the sights and sounds of the many birds including Bell Miners (*Manorina melanophrys*), finches, fantails and wrens. — **Toni Hess**

Above left: Bunya Pine **Above:** Giant Stinging Tree, *Dendronicide excelsa*

Look out for: Whiptail Wallabies (*Macropus parryi*) hopping through the forest; Yellow-tailed Black Cockatoos (*Calyptorhynchus funereus*) screeching above.

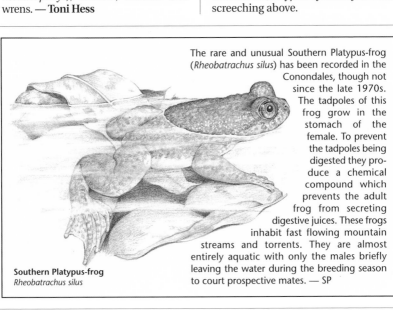

The rare and unusual Southern Platypus-frog (*Rheobatrachus silus*) has been recorded in the Conondales, though not since the late 1970s. The tadpoles of this frog grow in the stomach of the female. To prevent the tadpoles being digested they pro-duce a chemical compound which prevents the adult frog from secreting digestive juices. These frogs inhabit fast flowing mountain streams and torrents. They are almost entirely aquatic with only the males briefly leaving the water during the breeding season to court prospective mates. — SP

Southern Platypus-frog
Rheobatrachus silus

CROW'S NEST FALLS NATIONAL PARK

Crow's Nest Falls National Park is a 489 ha park about 6 km south-east of the township of Crow's Nest. Crow's Nest Creek flows through the park joining Perseverance and Cressbrook Creeks to the east.

The park terrain is very rugged and dominated by domed granite outcrops. The granites of the area contain large amounts of the mineral feldspar which sparkles in the sunlight, hence the name of the principal gorge of the park, the Valley of Diamonds.

Vegetation in the park consists mainly of eucalypt forest interspersed with several *Angophora* species. The creeks are lined with callistemons, casuarinas and Swamp Box (*Lophostemon suaveolens*).

LOCALITY GUIDE

Location: 120 km west of Brisbane GPO; 2 hours' driving.

Access: Via Esk.

Facilities: Picnic areas, tables, shelter sheds, toilets, barbecues, water; camping; walking tracks; lookouts; caravan access.

Restrictions: No domestic animals.

Crow's Nest has a rich bird life and common species include the Peaceful Dove (*Geopelia striata*), Pale-headed Rosella (*Platycercus adscitus*), Striated Pardalote (*Pardalotus striatus*) and Noisy Miner (*Manorina melanocephala*). Wedge-tailed Eagles (*Aquila audax*) ride the thermals while

Opposite: Crow's Nest Falls **Inset:** Brush-tailed Rock Wallaby *Petrogale penicillata* **Above:** Crow's Nest Creek

Peregrine Falcons (*Falco peregrinus*) soar above the gorges. Patient observers may see the cryptic Red-chested Button-quail (*Turnix pyrrhothorax*) searching the leaf letter or the White-throated Nightjar (*Eurostopodus mystacalis*) hawking the evening air.

A nocturnal ramble will reveal many of the park's animals such as the Rufous Bettong (*Aepyprymnus rufescens*), Greater Glider (*Petauroides volans*), Sugar Glider (*Petaurus breviceps*) and Feathertail Glider (*Acrobates pygmaeus*). Brush-tailed Rock Wallabies (*Petrogale penicillata*) inhabit the rocky outcrops and the solitary Yellow-bellied Sheathtail-bat (*Saccolaimus flaviventris*) patrols the night sky.

Reptiles are well represented in the park with Lace Monitors (*Varanus varius*) and Eastern Water Dragons (*Physignathus lesueurii*) commonly encountered. Small, retiring reptiles include the Thick-tailed Gecko (*Nephrurus milii*), Burton's Snake Lizard (*Lialis burtonis*), the Bandy Bandy (*Vermicella annulata*) and Red-naped Snake (*Furina diadema*). — **Rod Hobson**

Look out for: Brown Goshawk (*Accipiter fasciatus*) soaring over forest canopy; Emerald-spotted Treefrogs (*Litoria peronii*) calling from creekside vegetation at night.

Upper slopes, Valley of the Diamonds

GOOMBURRA STATE FOREST

Goomburra State Forest nestles in the foothills of the Great Dividing Range and borders Main Range National Park. The forest is one of the best locations to discover the northern section of the Scenic Rim and the Central Eastern Rainforest Reserves of Australia's World Heritage Area (see p. 163).

The extensive network of walking trails leads visitors through some of Queensland's most spectacular forests to panoramic views from lookouts perched high on the Great Dividing Range.

The topography of Goomburra ranges from the alluvial flats adjacent to Dalrymple Creek, through gently undulating slopes of the lower foothills, to the precipice of the eastern face of the Great Dividing Range. The diverse topography supports an enor-

mous range of vegetation communities, including eucalypt forests and rainforests, which in turn provide habitats for a similarly rich fauna.

North from Mt Castle lookout track

A number of interesting eucalypt species are found at Goomburra — the New England Blackbutt (*Eucalyptus andrewsii campanulata*) and Manna Gum (*E. nobilis*). New England Blackbutt has a limited distribution in Queensland, occurring only at higher altitudes around the Warwick and Stanthorpe areas and as sporadic, isolated patches elsewhere.

Manna Gum has an extensive distribution within south-eastern Australia but is not common in Queensland. The Manna Gum forests at Goomburra represent the northernmost occurrence of this species. Manna Gum can be easily identified by the long ribbons of bark which drape from the tree while the bark is being shed.

Goomburra State Forest also contains some of Queensland's most beautiful rainforest. Exploring these areas can reveal magnificent examples of Hoop Pine (*Araucaria cunninghamii*), the gnarled appearance of which gives an indication of the age of these trees and their exposure to harsh climatic conditions. These rainforests gained World Heritage significance in 1995 when they were listed along with extensive sections of the Scenic Rim as part of the Central Eastern Rainforest Reserves of Australia's World Heritage.

The Tiger Quoll (*Dasyurus maculatus*), also known as the Spotted-tail Quoll, is declining rapidly due to epidemics, habitat destruction, poisoning, trapping, shooting and introduced predators. Each quoll needs a large territory covering hundreds of hectares and remote mountainous areas like Goomburra and the rest of Main Range may be the quoll's best hope for survival. Quolls are the largest marsupial carnivores in mainland Australia and will eat small ground mammals, birds and reptiles. Quolls can sometimes be seen in bush camping grounds foraging for scraps.

Top: Dalrymple Creek **Centre:** Dingo **Opposite:** Dalrymple Creek

The fauna of Goomburra makes a trip to the area worth the effort. Some of Australia's most recognisable wildlife icons such as the Koala (*Phascolarctos cinereus*), Platypus (*Ornithorhynchus anatinus*) and the Echidna (*Tachyglossus aculeatus*) can be seen in the area. The Tiger Quoll (*Dasyurus maculatus*) has been recorded at Goomburra as has the Feathertail Glider (*Acrobates pygmaeus*) and the Mountain Brushtail Possum (*Trichosurus caninus*). Goomburra is also noted for its birds.

Like many state forests, Goomburra is managed for multiple use including timber and honey production, recreation, grazing, conservation and water catchment protection. — **Mark Peacock**

Look out for: Glossy Black Cockatoos (*Calyptorhynchus lathami*) screeching throughout the forest and the ringing of Bell Miners (*Manorina melanophrys*) in and around the camping area.

Tiger Quoll *Dasyurus maculatus*

KONDALILLA FALLS NATIONAL PARK

Kondalilla Falls National Park has an area of 327 ha and protects dry and wet eucalypt forest and subtropical rainforest.

The entrance to the park leads the visitor through an open picnic ground surrounded by eucalypts, particularly Tallowwood (*Eucalyptus microcorys*) and Grey Gum (*E. propinqua*).This small national park contains remnants of the once extensive forests of the Blackall Range. The Southern Platypus -frog (*Rheobatrachus silus*) was discovered here in the 1970s (see p. 146).

Very large Tallowwoods, Brush Box (*Lophostemon confertus*) and Flooded Gum (*E. grandis*) and stands of Piccabeen Palm (*Archontophoenix cunninghamiana*) can be seen along the track to the waterfall. Closer to the falls, the vegetation changes to a dry eucalypt forest and woodland with numerous grass trees growing out of the steep slopes.

The rock pool near the top of the waterfall is popular for swimming but be cautious. Do not dive or jump into the pool because there are submerged rocks.

From the top of the falls, which drop about 80 m, there are magnificent

LOCALITY GUIDE

Location: 110 km north of Brisbane GPO; 1 hours'driving.

Access: Via the Mapleton-Maleny Road from Landsborough, Kenilworth or Nambour.

Facilities: Picnic areas, barbecues, shelter, toilets, water; walking tracks; lookouts; disabled access.

Restrictions: No domestic animals; no camping.

views to the north-west down the Skene Creek Valley. Emergent Hoop Pines (*Araucaria cunninghamii*) pierce the rainforest canopy.

At least 107 species of birds have been recorded at Kondalilla and in nearby Mapleton National Park. The loud call of the Wompoo Fruit-dove (*Ptilinopus magnificus*) is one of the most distinctive sounds of the park. — **Stephen Poole**

Look out for: Multi-coloured Superb Fruit-doves (*P. superbus*) in the canopy; Log-runners (*Orthonyx temminckii*) scratching in forest litter.

Opposite: Palm forest **Above:** Base of falls

LAMINGTON NATIONAL PARK

Much has been written about Lamington which is perhaps the best known national park in Queensland and one of the most heavily visited in the region.

On weekends and holidays it can be hard to find picnic space in the two main entrance areas at Green Mountains and Binna Burra.

From these two access points, more than 150 km of well developed walking trails of varying grades and lengths fan out through the park.

All the graded walks can be undertaken by people with reasonable fitness levels. Additional bush tracks provide further scope for exploration by seasoned bushwalkers.

The original track system was constructed in the 1930s during the Depression. The park's high visitation levels mean that people "traffic" on some of the walking tracks is high but because of the steep topography of the park there is limited scope for spreading the load over new trail areas.

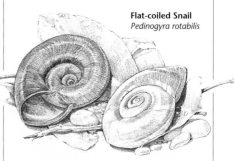

Flat-coiled Snail
Pedinogyra rotabilis

Lamington forms part of the huge shield volcano that in prehistory was centred on Mt Warning in northern New South Wales (see p. 95). With the McPherson, Tweed and Nightcap Ranges, the Lamington Plateau was part of volcano's massive caldera or rim. Erosion of the soft volcanic rocks has resulted in an extremely rugged terrain with a high plateau, many deep gorges and waterfalls.

Lamington has World Heritage listing and deserves its status. The park covers about 20,000 ha and contains the largest stands of sub-tropical rainforest and undisturbed warm temperate rainforest in Queensland.

Opposite: Near Coomera Falls

East from Surprise Rock

There is also dry rainforest, cool temperate rainforest, montane heath, and dry and wet eucalypt forests.

Much of Lamington is characterised by massive rainforest trees and spectacular, remote areas such as the Lost World and Valley of the Pines. Rainforest species include the Wheel of Fire (*Stenocarpus sinuatus*) and Flame Tree (*Brachychiton acerifolius*) which are highly visible during flowering because of their bright red or scarlet leaves. The Corkwood Tree (*Duboisia myoporoides*) is easily identified by the deeply fissured bark which has a cork texture.

The cool, temperate rainforest is generally found at elevations above 900 m. There are also considerable areas of wet eucalypt forest with large canopied trees such as Tallowwood (*Eucalyptus microcorys*), Flooded Gum (*E. grandis*), Sydney Blue Gum (*E. saligna*), Grey Gum (*E. biturbinata*) and New England Blackbutt (*E. andrewsii campanulata*). Other large trees include the Hoop Pine (*Araucaria cunninghamii*) and Brush Box (*Lophostemon confertus*).

An area of montane heath at Dave's Creek contains the mallee eucalypt, *Eucalyptus codonocarpa*, (see p. 23) The distinctive aroma of the Lemon-scented Tea-trees (*Leptospernum petersonii*) adds to the sensory experience of walking through this heath.

Dave's Creek is accessed from Binna Burra and is just south of the Ship Stern Range, high above the western side of Numinbah Valley. From here there are views of Springbrook, Numinbah and Mt Warning.

The main Border Walk track in the southern part of the park is one of the best places to see stands of ancient Antarctic Beech (*Nothofagus moorei*) (see opposite and p. 15). This area is often shrouded in cloud from which the trees draw moisture.

The Antarctic Beech (*Nothofagus moorei*) grows at only a few locations in South-east Queensland and the McPherson Ranges of the Queensland/New South Wales border are the most northerly occurrence of this tree. These ancient and multi-stemmed trees — some of which are believed to be more than 1000-years-old — grow at the highest elevations in cool temperate rainforest. For many years it was thought that the Antarctic Beech was dying out as very few, if any, young saplings had been observed. Although this situation has now changed, it may be that the effects of global warming will eventually cause their extinction in this region. In which case, southern Queensland will lose not one, but two plant species — the Beech Orchid (*Dendrobium falcorostrum*) grows only on branches of the Antarctic Beech.

Top: Lianes **Above:** Antarctic Beech

Top: Coomera Falls **Centre:** Fig roots
Above: Hidden Valley

Lamington is noted not only for its forests but also for its abundant wildlife. The Lamington Blue Spiny Crayfish (*Euastacus sulcatus*), which also occurs at Springbrook National Park (see p. 181). can be seen in rock pools at the base of waterfalls and sometimes on walking trails during summer months. When confronted by bushwalkers, the crayfish rears up and hisses while scrambling backwards.

Brilliantly coloured birds such as the Regent Bowerbird (*Sericulus chrysocephalus*), Crimson Rosella (*Platycercus elegans*), Rainbow Lorikeet (*Trichoglossus haematodus*) and King Parrot (*Alisteris scapularis*) are among the best known attractions of the park and some will even frequent the crowded picnic areas. Other significant species include the Rufous Scrub Bird (*Atrichornis rufescens*), Eastern Bristlebird (*Dasyornis brachypterus*), Albert's Lyrebird (*Menura alberti*) and Powerful Owl (*Ninox strenua*).

The Long-finned Eel (*Anguilla reinhardtii*) is found in many of the rock pools and creeks in Lamington and sometimes even in small pools formed by rain. These freshwater fish, which grow up to 2 m, are born in the oceans and return there to breed. The eels are capable of moving considerable distances over land on rainy nights. It is sobering to encounter these eels at more than 700 m elevation, knowing that they have been born in salt water more than 100 km away.

The Green Mountains and Binna Burra entrance areas are best experienced in the early morning or late afternoon when the bulk of visitors has left. At these times, it is possible to see Red-necked Pademelons (*Thylogale thetis*) feeding on the grass. — **Stephen Poole.**

Long-finned Eel *Anguilla reinhardtii*

Look out for: Giant Pill Bugs (*Sphaerillo grossus*) which curl into an "armoured" ball when disturbed; lots of different fungi growing on logs — some are luminescent.

Top left: Tree ferns **Above:** Egg Rock

MAIN RANGE NATIONAL PARK

Main Range National Park covers 18,400 ha of the Great Dividing Range stretching from Wilson's Peak near the Boonah border gate through to Mt Mistake, south-west of Laidley.

Most of Main Rain National Park is undeveloped and it offers a true wilderness experience for those willing to venture away from the picnic grounds and short walking tracks. The rugged escarpments of Main Range have a beauty that rivals any of the better known and more accessible reserves around the Greater Brisbane Region. The park is included on the World Heritage listing under the Central Eastern Rainforests Reserves of Australia nomination (see p. 151).

The magnificence of the park is best experienced in the early morning as the sunlight defines the craggy recesses of the range; or in the late afternoon when the views to the east are bathed in brilliant light but the observer stands in a cool, dark remnant of the vast forests that once stretched as far as the eye could see.

The national park protects open eucalypt forest, wet and dry sub-tropical rainforest and montane heath. The

LOCALITY GUIDE

Location: 116 km south-west of Brisbane GPO; 1.5 - 2 hours' driving.

Access: The Cunningham Highway runs through the middle of the park. Spicer's Gap can only be reached by conventional vehicles via the Cunningham Highway and Moogerah Dam Road during dry weather.

Facilities: Picnic areas, toilets, tables, shelter sheds, barbecues, water; walking tracks and hiking trails; lookouts; camping; disabled access; caravan access.

Restrictions: No domestic animals; no mountain bikes; advance permits for remote camping required.

park is not contiguous but comprises a number of sections, buffered by state forests, the biggest of which is Goomburra (see p. 151).

Macleay's Swallowtail Butterfly
Graphium macleayanus

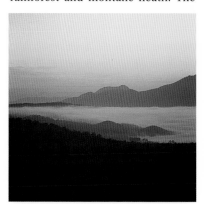

Opposite: Mt Mitchell **Above:** Teviot section

Above: Near Teviot Gap **Right:** Spicer's Gap,
Grass trees (*Xanthorrhoea glauca*)

Mt Superbus, in the southern section of the park is the highest peak in Southeast Queensland and second only to Mt Bartle-Frere in the far north of the state. (Although Mt Barney is the highest free-standing mountain, see p. 173).

As with other near wilderness areas, many rare and restricted species of plants and animals are found within the boundaries of the park. Rocky areas throughout Main Range are home to the Brush-tailed Rock Wallaby (*Petrogale penicillata*). The beautiful Cunningham's Skink (*Egernia cunninghami*) is another inhabitant of the rocky outcrops.

CUNNINGHAM'S GAP

This is the most popular section of Main Range National Park as the Cunningham Highway passes through the Gap and provides easy access to camping and picnic areas and a number of walking tracks.

Most of the tracks begin near the memorial to Alan Cunningham who was the first European to locate the gap in the ranges in 1828.

From Cunningham's Gap, graded walking tracks lead to the east and west peaks of Mt Mitchell, Bare Rock, Morgan's Lookout, Mt Cordeaux and Gap Creek Falls. A short walk from the Gap leads to the Fassifern Valley Lookout.

Mt Mitchell from Mt Cordeaux, Spear Lily (*Doryanthes palmeri*) foreground

Mt Mitchell

Pelican Spiders, so-named for their remarkable resemblance to the bird, were first discovered as fossils in Baltic amber and only later were living examples found in Madagascar, Chile, New Zealand and Australia. These small spiders inhabit rainforests, wet eucalypt and Antarctic Beech forests where they spin a single thread of silk in shrubs or in dense moss.

Pelican Spider
Austrarchaea nodosa

The remoter areas of the park can be reached via numerous unmarked tracks. Bush campsites suitable for small numbers of people can also be found in these remote areas. Visitors to the remote parts of the park will require a high level of physical fitness, map reading skills and good bush sense. Bush campsites must be booked in advance through the park office at Cunningham's Gap.

Hoop Pine (*Araucaria cunninghamii*), Red Apple (*Acmena ingens*) easily identified by its large red fruits, the Giant Stinging Tree (*Dendronicide excelsa*), Scrub Ironwood (*Austromyrtus acmenoides*), Native Frangipanni (*Hymenosporum flavum*), Brush Box (*Lophostemon confertus*) and Piccabeen Palms (*Archontophoenix cunninghamiana*) grow in the rainforest around the Gap.

Near the top of Mt Cordeaux, giant Spear Lillies (*Doryanthes palmeri*) can be found growing in rocky exposed areas. These rare and spectacular plants, with sword-like leaves up to 3 metres long, burst into spring colour with spikes of red flowers up to a metre long.

Near Cunningham's Gap listen for Bellbirds (*Manorina melanophrys*) in sheltered gullies. The birds can even be heard on the Cunningham Highway as you climb towards the Gap — so turn off the radio and wind down the windows.

Top left: Mt Wilson **Above:** Rainforest near Teviot Gap

SPICER'S GAP

This is another popular area accessed from Spicer's Gap Road which branches off the Cunningham Highway 5 km west of Aratula.

The historic Spicer's Gap road was first cleared in 1847 as a dray route between early settlements on the Darling Downs and the Moreton Bay colony (later named Brisbane). Today the Spicer's Gap area abounds in history from these early days. Short walking tracks lead to the Pioneer Graves, Moss' Well and Governor's Chair lookout. A walking track along a 1.6 km section of the historic road, west of Governor's Chair, has signs to interpret key engineering features. — **Stephen Poole**

Look out for: Lots of noisy cicadas in the summertime; an occasional glimpse of Albert's Lyrebird (*Menura albertii*) on walking tracks.

Opposite: Landslip, Teviot Falls

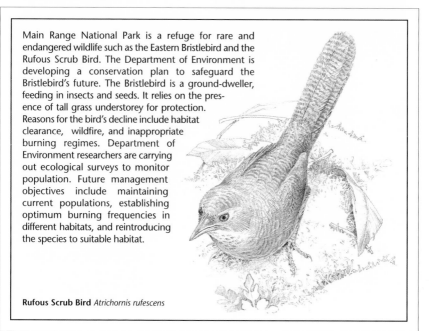

Main Range National Park is a refuge for rare and endangered wildlife such as the Eastern Bristlebird and the Rufous Scrub Bird. The Department of Environment is developing a conservation plan to safeguard the Bristlebird's future. The Bristlebird is a ground-dweller, feeding in insects and seeds. It relies on the presence of tall grass understorey for protection. Reasons for the bird's decline include habitat clearance, wildfire, and inappropriate burning regimes. Department of Environment researchers are carrying out ecological surveys to monitor population. Future management objectives include maintaining current populations, establishing optimum burning frequencies in different habitats, and reintroducing the species to suitable habitat.

Rufous Scrub Bird *Atrichornis rufescens*

MARY CAIRNCROSS PARK

This small park, only 52 ha in area, is administered by Caloundra City Council. The land was donated by the Thynne family in memory of their relative Mary Cairncross in 1941 as a scenic reserve for preservation and conservation.

This generous act of foresight has enabled a superb stand of sub-tropical rainforest to survive while other similar areas on the Blackall Ranges and around Maleny have been lost or degraded.

The park is rich in wildlife including rarities such as the Eastern Pygmy Possum (*Cercartetus nanus*) and the Burrowing Skink (*Coeranoscincus reticulatus*). The daylight hours favour birds and reptiles. Small birds, among them Pale-yellow Robins (*Tregellasia capito*), through to the larger pigeons, parrots and raptors, can be seen throughout the park. Mammals, reptiles and frogs are less obvious but include the Brown Antechinus (*Antechinus stuartii*), Tyron's Skink (*Eulamprus murrayi*) and Whistling Treefrog (*Litoria verreauxii*).

However it is the vegetation which makes a visit to this park worthwhile and as an added bonus, many of the trees contain labels to assist interpretation and education.

There are giant Strangler Figs (*Ficus watkinsiana*), Black Bean trees (*Castanospermum australis*) with their large signature seed cases, and many Native Tamarind (*Diploglottis australe*) trees. Red Cedar (*Toona ciliata*), Yellow Carabeen (*Sloanea woollsii*), Brown Tulip Oak (*Argyrodendron trifoliolatum*), Black Apple (*Planchonella ciliata*), and White Beech (*Gmelina leichhardtii*) also grow here.

The walking trail has a couple of loops

which return to the entrance to the park. At one point the trail passes an extensive forest of Piccabeen Palms (*Archontophoenix cunninghamiana*) which straddles a slow-moving creek.

Ignore the sounds of the outside world which occasionally intrude and try to spend at least two hours in the forest slowly walking and listening — preferably early or late in the day as the park is popular with coach tours and school groups — **Stephen Poole.**

Look out for: Mountain Brushtail Possums (*Trichosurus caninus*) in the evening; King Parrots (*Alisteris scapularis*) perched in the trees surrounding picnic area.

In the Southern Hemisphere, lianes and vines grow in an anti-clockwise spiral. This is the opposite to the Northern Hemisphere where they grow clockwise. This phenomenon is known as the Coriolis Effect and also influences the way water runs down the plug hole  — again clockwise in the Northern Hemisphere and anti-clockwise in the Southern Hemisphere. — SP

MT BARNEY NATIONAL PARK

On clear days, the prominent peaks of Mt Barney and nearby Mt Lindesay can be seen from high rise offices in the Brisbane Central Business district, even though it takes about 1.5 hours by car to get there.

Mt Barney is the highest free-standing peak (1357 m) in South-east Queensland and also one of the most easily recognised landmarks. Along with Mt Lindesay, Mt May, Mt Maroon, Mt Ballow, Mt Ernest and Durramlee, Gwyala and Double Peaks, Mt Barney is a remnant of the Focal Peak Shield Volcano, which like the Tweed Volcano (see p. 95), gave rise to the rugged landscapes of the McPherson Ranges.

The precipitous terrain of Mt Barney, and other national parks of the Scenic Rim (see p. 4), works in favour of conservation and assists plant and animal species to survive, either

Opposite: East and West Peaks from Mt Barney Gorge

LOCALITY GUIDE

Location: 130 km south-west of Brisbane GPO; 1.5 hours' driving.

Access: Off Mt Lindesay Highway, 20 km south-west of Rathdowney.

Facilities: Picnic areas, toilets; camping; walking tracks and hiking trails.

Restrictions: No domestic animals.

through physical remoteness or by acting as natural barriers.

Mt Barney National Park incorporates several different habitats including dry eucalypt forest which covers most of the area, warm temperate rainforest and at higher elevations, montane heath and cool temperate rainforest.

Most of the harshly beautiful terrain of the park is undeveloped. The often dense scrubby conditions and precipitous rock faces are suitable only for experienced bushwalkers. First-time walkers at Mt Barney should consult with the rangers and carry a good topographic map, compass, water and suitable warm clothing.

Those who want to sample the beauty of the area, but not tackle a challenging walk, can follow the track from the carpark at Yellowpinch to the Lower Portals on Barney Creek. This 3.7 km walk takes about an hour each way. The walk to Mt Lindesay is very difficult and inaccessible for most people.

It is also possible to get a feel for the area by driving around it. One option is to start at Rathdowney, drive to the Yellowpinch carpark, then backtrack and follow the Mt Lindesay Highway to the Border Gate. Proceed to Woodenbong in New South Wales and then follow the Mt Lindesay Highway west, turn off onto White Swamp Road and travel to the Boonah Border Gate. Drive past Maroon Dam, back to Rathdowney or Boonah. Allow four hours for a leisurely drive with plenty of short viewing stops.

The rocky terrain of the Mt Barney area is ideal habitat for the Brush-tailed Rock-wallaby (*Petrogale penicillata*). Populations of this species are in decline. The wallaby inhabits the steep, exposed slopes which often contain rocky shelters near patches of mature grasses. Another restricted species is the Eastern Pygmy Possum

Pygmy Possum

(*Cercartetus nanus*) which is found in the montane heath at Mt Barney and Lamington. The possum depends on pollen and nectar from the heath plants for its survival. — **Stephen Poole**

Look out for: Platypus (*Ornithorhynchus anatinus*) in Mt Barney Creek; colourful honeyeaters and robins.

Above: From the Lower Portals **Opposite:** Mt Barney Creek

RAVENSBOURNE NATIONAL PARK

Ravensbourne National Park on a spur of the Great Dividing Range is directly east of the township of Hampton. The city of Toowoomba lies 59 km south-west (by bitumen road).

The park was declared in 1922 and covers 100 ha. Its main attraction is lush rainforest with a profusion of plant species. Sydney Blue Gum (*Eucalyptus saligna*), Red Cedar (*Toona ciliata*), Moreton Bay Fig (*Ficus macrophylla*) and Black Bean (*Castanospermum australe*) are some of the dominant canopy species which thrive on the red basaltic soils of the park.

These trees provide a moist, shaded atmosphere below for a riot of ferns, fungi and mosses. They also support epiphytic orchids, Crow's Nest (*Asplenium australasicum*), Staghorn (*Platycerium superbum*) and Elkhorn (*P. bifurcatum*) ferns. Groves of Piccabeen Palm (*Archontophoenix cunninghamiana*) link Palm and Buaraba Creeks, the two waterways crossing the park.

Fruiting rainforest trees provide food for a variety of birds which can be seen by the patient observer. The Wompoo Fruit-dove (*Ptilinopus magnificus*), Rose-crowned Fruit-dove (*P. regina*) and Superb Fruit-dove (*P. superbus*) all inhabit the tree-tops. Their forest floor relative, the dainty Emerald Dove (*Chalcophaps indica*) can often be flushed from the trackside where it feeds on seeds, berries and fallen fruit knocked down by its arboreal cousins. Noisy Pittas (*Pitta versicolor*) and Australian Brush-turkeys (*Alectura lathami*) also feed along the tracks. Topknot Pigeons (*Lopholaimus antarcticus*), Regent (*Sericulus chrysocephalus*) and Satin (*Ptilinorhynchus violaceus*) Bowerbirds and Paradise Riflebirds (*Ptiloris paradiseus*) occupy

L O C A L I T Y   G U I D E

Location: 110 km west of Brisbane GPO; 1.5 hours' driving.

Access: Via Esk on the Esk-Hampton Road.

Facilities: Picnic areas, shelter; toilets; water; walking tracks; lookouts; disabled access.

Restrictions: No domestic animals; no camping.

the upper levels of the rainforest.

Birds of prey such as the Grey Goshawk (*Accipiter novaehollandiae*) and the Peregrine Falcon (*Falco peregrinus*) and their nocturnal counterparts, the Sooty (*Tyto tenebricosa*) and Powerful Owls (*Ninox strenua*) occupy the upper avian hierarchy of the park. The rare Black-breasted Button Quail (*Turnix melanogaster*) forages in patches of dry vine scrub along the park edges.

Opposite: Piccabeen Palm in flower
Above: Water Vine

The mammals at Ravensbourne are rarely seen — most are nocturnal. A strong spotlight is needed to reveal some of the park's more retiring denizens such as the ubiquitous Bush Rat (*Rattus fuscipes*) and its relative, the arboreal Fawn-footed Melomys (*Melomys cervinipes*).

The park's marsupials include the Yellow-footed Antechinus (*Antechinus flavipes*), Sugar Glider (*Petaurus breviceps*), Common Ringtail Possum (*Pseudocheirus peregrinus*), Mountain Brushtail Possum (*Trichosurus caninus*) and Common Brushtail Possum (*T. vulpecula*). Red-necked Wallabies (*Macropus rufogriseus*), Swamp Wallabies (*Wallabia bicolor*) and Greater Gliders (*Petauroides volans*) inhabit the more open forests.

The rare Eastern Chestnut Mouse (*Pseudomys gracilicaudatus*) is found in adjoining grasslands and a large colony of Eastern Horseshoe Bats (*Rhinolophus megaphyllus*) occupy the cave system on the eastern edge of the park.

Ravensbourne has a healthy reptile fauna including several large snakes such as the Eastern Brown Snake (*Pseudonaja textilis*), Red-bellied Black Snake (*Pseudechis porphyriacus*), Common Tree Snake (*Dendrelaphis punctulatus*) and the Carpet Python (*Morelia spilota variegata*). Several smaller snakes and many lizards inhabit the litter of the forest floor. The largest lizard in the park is the Lace Monitor (*Varanus varius*), while the large Pink-tongued Skink (*Hemisphaeriodon gerrardii*) forages the forest floor by night.

Notable amphibians include the Green-thighed Frog (*Litoria brevipalmata*). — **Rod Hobson**

Look out for: Scarlet-sided Pobblebonks (*Limnodynastes terraereginae*) on tracks after summer rains. Dollarbirds (*Eurystomas orientalis*) hawking insects above the forest canopy.

SPRINGBROOK NATIONAL PARK

The Springbrook plateau varies in altitude from 600 m above sea level to more than 1000 m. Like Lamington, Springbrook is part of the northern rim of the ancient Tweed Shield Volcano (see p. 95).

The first national parks at Springbrook were gazetted in 1937 and 1940 and the plateau originally included several smaller parks which, although still separated, have now been incorporated into Springbrook National Park.

The new park covers about 3000 ha and the main sections are: Springbrook, Natural Bridge and Mt Cougal. The park protects wet and dry eucalypt forests, warm temperate and

LOCALITY GUIDE

Location: 110-130 km south of Brisbane GPO; 1.5 hours' driving.

Access: Via Nerang, Currumbin or Mudgeeraba Roads.

Facilities: Picnic areas; camping; walking tracks, interpretation centres.

Restrictions: No domestic animals; no fires.

sub-tropical rainforests and at its highest altitudes, cool temperate rainforest. It contains the highest number of endemic, rare, endangered and noteworthy species of any reserve in the Greater Brisbane region.

Opposite: Eucalypt forest against a backdrop of sub-tropical rainforest

Above: View towards Little Nerang Creek **Opposite:** Antarctic Beech (*Nothofagus moorei*)

The tall dry eucalypt forests of the Springbrook Plateau are dominated by Blue Mountains Ash (*Eucalyptus oreades*) and New England Blackbutt (*E. andrewsii campanulata*) which are at, or near, their most northerly limit here. Flooded Gum (*E. grandis*) is a notable species of the wet eucalypt forest which has an understorey of palms and tree ferns. Sub-tropical rainforest can be found at the edges of the plateau in sheltered gorges and gullies.

In the Springbrook section of the park, one of the most popular spots is Purlingbrook Falls which drops some 106 m into the gorge below. From here visitors can walk through eucalypt forest and sub-tropical rainforest. A giant Red Cedar (*Toona ciliata*) survives near the base of the falls. On summer evenings

Tiger Snake *Notechis scutatus*

Inset: Right, Boobook Owls *Ninox novaeseelandiae*

Above: Purlingbrook Falls

walkers might encounter the Great Barred-frog (*Mixophyes fasciolatus*) and the Australian Marsupial Frog (*Assa darlingtoni*).

One of the best walks in the park is the Warrie Circuit which winds 17 km down spectacular bluffs and past numerous waterfalls. It is possible to walk behind Twin, Rainbow and Black-fellow Falls but be careful along cliff edges and on wet rock surfaces which can be slippery. Stands of Brush Box (*Lophostemon confertus*) below the cliffs give way to sub-tropical rainforest.

The loud call of the Yellow-tailed Black Cockatoo (*Calyptorhynchus funereus*) is one of the most distinctive sounds of the park and the birds can also be heard crunching on casuarina seeds. The Wedge-tailed Eagle (*Aquila audax*) can be seen from cliff vantage points. Regent and Satin Bowerbirds (*Sericulus chrysocephalus* and *Ptilinorhynchus violaceus*) are also common at Springbrook along with the King Parrot (*Alisteris scapularis*), Crimson Rosella (*Platycercus elegans*) and the secretive Albert's Lyrebird (*Menura alberti*).

NATURAL BRIDGE

LOCALITY GUIDE

Location: 110 km south of Brisbane GPO; 1.5 hours' driving.

Access: Numinbah Valley Road.

Facilities: Picnic area, toilets.

Restrictions: No domestic animals; no camping.

The Natural Bridge section of Springbrook is located in the scenic Numinbah Valley and stretches from the base of the valley east to the highest point on the Springbrook Plateau.

Sub-tropical rainforest covers most of the park changing as elevation increases to cool temperate rainforest near the Springbrook Repeater Station and Best of All Lookout.

Most of the park is inaccessible and rugged but two major points of interest are only a short walk from the carparks. The park takes its name from an unusual geological feature — a rock arch, thought to have been created by a stream breaching the roof of an underground cavern. The breach was probably made by the scouring action of rocks caught in a sink hole in the basalt river bed.

Inside the cave is one of the best places to see glowworms in South-east Queensland. The glowworms appear as tiny pricks of light on the roof of the cave. Glowworms are not true worms but the larvae of the a fungus gnat (*Arachnocampa flava*) — see opposite. The best time to see the glowworm display is after dark and they tend to be more prevalent after rain. Visitors should be careful not to touch the glowworms. Smoking is not permitted in the caves.

The Springbrook end of the park rises to 1170 m above sea level and from Best of All Lookout there are panoramic views south-east to Murwillumbah and Byron Bay and the Mt Cougal section. About 30 m before the lookout is a small stand of Antarctic Beech (*Nothofagus moorei*) — see p.159.

Glowworms can be found in moist, sheltered areas, such as embankments, overhangs or caves. Although they are called worms, they are actually the larvae of a fungus gnat which lays its eggs on the roof of caves. When the larvae hatch they spin fine threads of silk up to a few centimetres long. The dangling silk threads are covered with small droplets of a sticky, glue-like substance which catch small insects attracted to the luminescent light given off by a special organ on the abdomen of the larvae. The light is caused by a chemical reaction in the body of the larvae and is similar to that obtained by mixing the chemicals in commercially available "glo-lights".

Glowworm

MT COUGAL

LOCALITY GUIDE

Location: 100 km south of Brisbane GPO; 1.5 hours' driving.

Access: Cougal's Cascades via Currumbin Valley Road, off Gold Coast Highway; east peak of Mt Cougal via Garden of Eden Road, Tomewin.

Facilities: Picnic ground, water, toilets; walking tracks; lookout; disabled access.

Restrictions: No domestic animals; no camping; no open fires; no trail bikes.

The rugged twin peaks of east and west Mt Cougal dominate the landscape of this densely vegetated section of Springbrook National Park.

Sub-tropical rainforest covers most of the park, encircling the peaks, although the northern and western slopes support tall eucalypt forest. The giant Spear Lilly (*Doryanthes palmeri*) grows on the east peak of Mt Cougal.

The twin peaks are usually reached only by experienced bushwalkers and development within the park is limited to the Cougal's Cascades site. Here a graded track from the picnic ground winds its way past the cascades (the headwaters of Currumbin Creek) to the remains of an old sawmill which in the 1940s produced packing crate timber for local banana farmers from Flooded Gum (*Eucalyptus grandis*) and Blue Fig (*Elaeocarpus grandis*). — **Stephen Poole**

Look out for: Small marsupials like the Antechinus (*Antechinus stuartii*) during summer months; luminescent fungi amongst the foliage after dark.

Above: Giant Stinging Tree *Dendronicide excelsa*

The Yugambeh Aboriginal people have a long and continuing relationship with the Mt Cougal area. One of their legends describes how two prized hunting dogs belonging to Gwayla, a hunter from long ago, were killed. The dogs were buried under the twin peaks of Mt Cougal which were then known as Ningeroongun and Barrajanda in their memory.

PLANNING FOR WILDLIFE IN
THE GREATER BRISBANE REGION

Brisbane, with its range of habitats from the Bay coral reefs and seagrass meadows through the extensive creek systems to the Mt Glorious rainforests, could be called the "Wildlife Capital of Australia". But for animals to survive both within and outside protected areas such as Parks and Reserves and to give us a sense of "wild places" in our suburbs, we all have to act in our own localities.

Scarlet-sided Pobblebonk (*Limnodynastes terraereginae*)

To ensure that wildlife remains a part of our landscape we need to plan now for the provision, protection and enhancement of habitats, corridors, and vegetation remnants. Too many important wild places are being destroyed by the relentless progression of urbanisation and agricultural development.

People interested in the conservation of individual plants and animals need to consider the conservation of habitat throughout South-east Queensland. In urban areas this habitat is often an isolated piece of bushland that is degraded by weed invasion and neglect. By conducting weed control programs, providing connective corridors, planting additional species and providing animal homesites such as nest boxes, these areas can be improved for wildlife. Remnant areas and corridors are often the only wild places where many species of native plants and animals are able to survive.

Lycid Beetle (*Metriorrhynchus sp.*)

It is up to us to ensure the continued provision of food, shelter and home-sites for our wildlife.

How To Do It

There are four basic ways that wildlife habitats can be protected and enhanced — replanting, regenerating, creation or enhancement of corridors and garden support.

Replanting

Replanting is the provision of a new habitat and may provide a corridor or rehabilitate a degraded or cleared piece of land. Depending on the type and style of the work to be undertaken, different methods of approach are needed. If only a few trees are required to fill a gap an individual or family could plant them. If rehabilitation of a creek system is required many people would be involved and public meetings and organisation may be required. Essential to any replanting scheme is the use of the right plants, planted in the right place for the right reasons, otherwise success, from a wildlife perspective, will not be achieved.

Recent increases in the number and variety of bird and butterfly sightings in Brisbane as measured by projects such as Nature Search are directly linked to the increase of plantings of appropriate native trees and shrubs in backyards and neighbourhoods providing shelter and food.

Regeneration

Bush regeneration is the rehabilitation of bush from a weed-infested or otherwise degraded plant community to a healthy community composed of Australian plants suitable for native wildlife. Many urban landholders use their local bush remnant as a convenient rubbish tip. This increases the incidence of exotic weeds and encourages others to dump more. In her book *Bringing Back the Bush*, Joan Bradley outlined the following system of weed control:

Weeding a little at a time from the bush towards the weeds takes the pressure off the natives under favourable conditions. Native seeds and spores are ready in the ground, and the natural environment favours the plants that have evolved in it. The balance is tipped back towards regeneration. Keep it that way, by always working where the strongest area of bush meets the weakest weeds.

Common Brushtail Possums (*Trichosurus vulpecula*)

Other activities that can be undertaken to bring back the bush wildlife include:

• removing unsuitable native plants after natural regeneration;

• replanting with native species lost from the bushland or no longer naturally regenerating;

• planting with better suited native species in areas where the soil, water or fire regimes are irretrievably changed. A good example is the planting of stormwater drains and other artificial drainage lines with species that normally occur along creeks;

• total replanting, which may be necessary in highly disturbed areas.

The recipe for weed removal, replanting and planting of native species varies with every project and may vary even within the area regenerated. Successful rehabilitation, using the Bradley method as its basis, has been conducted along some of South-east Queensland's creeks and remnant bushland patches.

Publications such as Joan Bradley's *Bringing Back the Bush* and *Recovering Australian Landscapes*, by Robin Buchanan, detail work that can and should be undertaken to enhance local areas. If the area is large enough, a form of "wildness" can be created through regeneration.

Corridors

Corridors are valuable in nature conservation and general land management. A corridor is a relatively narrow piece of land linking two larger areas of bushland and which often contains essentially the same plant species and soil types. Corridors allow for movement of flora and fauna between larger habitat areas. They may vary from a single line of trees, providing a corridor for possums and magpies in an urban area, to wide and diverse connections between major national parks. Each has its own role to play.

Corridors can provide the means for animals and plants to repopulate an area as well as allowing young animals to disperse from the area where they originate. Plants need an opportunity for seed dispersal and for cross-pollination, allowing the exchange of genes between different plants, which ensures that variability and adaptability will persist within a wild population.

Corridors may remain after habitat destruction has encroached on a large area and almost divided it in two. Such remnant corridors need maintenance and perhaps regeneration to ensure their continued effectiveness.

When remnants of bushland are identified, their general conservation value should be established by identifying those which are likely to function as corridors. Of particular importance is their size, types of plant communities, occurrence of rare species, degree of weed invasion, the type of ownership and disturbance history. When this is compared with the distribution and abundance of animals on the regional scale, a determination can be made of each remnant in order to assess an area's value as a corridor.

Large, secure areas of native vegetation, such as parkland, completely separated by a narrow strip of degraded land, would be an obvious case for corridor re-establishment.

Golden-crowned Snake (*Cacophis squamulosus*)

The minimum width of the corridor needed is dependent upon the vegetation type and the species expected to use it, but in general, the wider the better. In heath vegetation a corridor width of 30 m could be adequate, but it should be at least 200 m for rainforest or tall eucalypt forest. Wider is better, and three to five rows of trees and shrubs provide a better shelter belt and corridor than a single row of trees. Existing corridors on public land can be enhanced by garden plantings of appropriate species on adjoining private land.

To be of greatest value for wildlife, the corridor should contain a diversity of plant species. These should include as many local species as possible, and contain all the strata found in the original vegetation.

Corridors have wider benefits which affect the general appearance and productivity of the landscape. They help to combat land degradation, by reducing erosion and salinity and provide shade and shelter for stock.

The Home Garden

Exotic species which grow in many of our gardens can be a problem for wildlife if they become established in natural areas. The home garden when appropriately developed can become a wildlife habitat. Local flora, including trees, small plants and herbage should be introduced so that natural vegetation communities are maintained.

Galah (*Cacatua roseicapilla*)

A small pond with various water depths in a home garden can provide a useful habitat for birds, frogs and other animals. Where space is restricted a bird bath can be of benefit to some species. Make sure the bath is well out of the reach of cats and inaccessible to cane toads. The bath or pond should receive shade and have some dense shrubs nearby to serve as a retreat for smaller birds. Perches should be placed near the bath to allow birds to look around before flying down to drink. The sound of dripping or falling water attracts birds.

Flowering plants in even a small garden can be an oasis, attracting butterflies and other insects, honeyeaters, lorikeets and possums. A variety of these plants can ensure that nectar is available all year, attracting insects which, in turn, will be food for insect-eating birds. Judicious additional feeding from bird feeders, with seeds or a sucrose-water mix, may encourage desirable species initially, but avoid making them dependent on handouts. Natural food sources are by far the safest and best.

Butterflies have specific needs and breeding of species such as the Lemon Migrant (*Catopsilia pomona*), skipper and crow butterflies can be ensured by planting native cassias, mat rushes and native figs.

Denser prickly shrubs can be planted to give protection for wrens and finches. Zebra finches, for example, often choose to nest in a prickly or thorny shrub or small tree, or in a dense creeper, such as Traveller's Joy (*Clematis* spp.), which a cat cannot climb.

Mature trees with hollows may be absent from many recently-planted and urban areas, but nesting boxes can be placed in trees to supply the nesting and living spaces for birds, bats and possums. Letting a couple of metres of garden along the back fence "return to nature" will also assist in litter accu-mulation and with the addition of hollow logs there will be an increase in abundance of small invertebrates, lizards and frogs.

Domestic pets can have a dramatic impact on wildlife. Cats are the greatest direct threat to smaller wildlife, especially in suburban areas, killing a range of animals including birds, small marsupials, reptiles, and even certain bats. Cats can be wide ranging and a cat-free garden will soon be invaded by a neighbourhood cat.

A successful control method is to obtain a small dog. Dogs don't climb and can be trained not to chase wildlife but will discourage cats entering the yard. Other deterrents may be effective in the short term, but the cat's hunting instinct will prevail in time. If you choose to own a cat, have it desexed, keep it well fed, housed inside at night, and tie a bell on its collar to give the wildlife some chance.

To protect bushland adjacent to homes, the movement of cats and dogs which forage in these areas should be discouraged. They can soon kill or disturb native animals to the point where they become locally extinct.

Whether the actions are small, one person "fauna-scaping" a backyard garden; or large, a community group rehabilitating a creek; the cumulative effect will be a better environment, a more diverse flora and fauna and Brisbane will remain the "Wildlife Capital of Australia". — **Peter J.M. Johnston and Pat Comben**

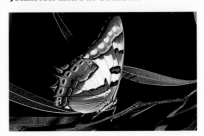

Tailed Emperor (*Polyura sempronius*)

ABOUT THE AUTHORS

Adrian Caneris is the Wildlife Conservation Officer with the Redland Shire Council. He was previously employed as a Wildlife Ranger with the Department of Environment's Nature Search field survey team. Adrian consistently talks to a wide range of community-based organisations on wildlife related topics. He founded the "Friends of White Rock" group and is currently the vice-president of the Wildlife Preservation Society of Queensland.

Pat Comben is Director of the Wildlife Preservation Society of Queensland. A former Queensland Environment Minister, Pat has a special interest in habitat modification and rehabilitation of degraded sites. He is presently working with the Brisbane City Council on a major rejuvenation project along Kedron Brook, a small north Brisbane waterway. As well as writing, much of his spare time is spent trying to see birds (particularly new types) in southern Queensland.

Greg Czechura works as a Senior Technician at the Queensland Museum. His qualifications include a Diploma in Teaching, an Associate Diploma in Applied Biology and a Bachelor of Education. His interests include birds of prey (about which he has published extensively), reptiles, amphibians and the history of the High Middle Ages. He is Queensland Area Coordinator and council member of the Australasian Raptor Association and is studying the distribution and status of the Red Goshawk, Australia's rarest bird of prey, in southern Queensland.

Shawn Delaney is a Special Projects Officer with the Department of Natural Resources. She has been employed in the field for about 12 years and holds various formal qualifications reflecting her interests in this area. In her spare time, Shawn can be found either exploring or planning her next wilderness escapade.

Martin Fingland is the Wildlife Manager of Gondwana Wildlife Sanctuary at South Bank. Before this, he worked at Brisbane Forest Park for seven years where he was in charge of the park's recreation and interpretation operations. Martin maintains an on-going relationship with Brisbane Forest Park regularly leading bushwalks out into the park's "wild places."

Toni Hess began work with the Department of Natural Resources in 1992 following completion of an Associate Diploma in Applied Science Forestry. Toni works in the field of forest interpretation and has been involved in interpretation of the Wet Tropics World Heritage Area and the state forests of the Conondale Range.

Rod Hobson has had a life-long interest in all aspects of natural history. He is especially interested in the less popular, obscure groups of animals. Although Rod has travelled extensively, he has found time to investigate and document the natural history of the Toowoomba region. Rod has also had field experience with a number of endangered species including the Black-breasted Button Quail, Red Goshawk and Water Mouse. His other interests include history and book collecting.

Peter J. M. Johnston, a Brisbane resident since 1970, has degrees in both Agricultural Science and Planning and has worked in the field of environmental conservation and natural resource management for 30 years in government and industry. As President of Greening Australia, a member of a number of community conservation groups and through his paid employment, he has lobbied for conservation of wildlife and associated habitat at the time of change in land use. Peter, an avid naturalist and birdwatcher, wrote *Grow Your Own Wildlife* in 1990. He has a loathing of feral cats and their impact on wildlife and is Australia's leading researcher on *Rubbathonghii*.

Tim Low is an environmental consultant and well-known nature writer and photographer. He is the author of five books including the prize-winning *Bush Tucker* and *Bush Medicine* and is a regular contributor to *Wildlife Australia, Nature Australia* and other magazines. Tim has conducted many flora and fauna surveys in the Greater Brisbane Region, on behalf of local councils and has worked at the Queensland Museum as an Interpretation Officer.

David Morgans has been the Manager of Brisbane Forest Park since 1898. He joined the park in 1986 as its first Planning Officer where his role was to facilitate visitor use and resource planning for all the "wild places" of Brisbane Forest Park. David is an Environmental Planner by profession, having previously worked as a consultant in private practice.

Mark Peacock is the Senior Officer responsible for state forest management in the Brisbane District of the Department of Natural Resources. The Brisbane District encompassing about 60 state forests over an area of 16,000 ha. These include some of the most heavily visited state forest in Australia. Mark is a professional forester who also has post-graduate qualifications in business administration. Mark is particularly interested in advancing sustainable natural area management, particularly in relation to achieving an appropriate balance between public access, resource utilisation and conservation.

Stephen Poole is an Environment Officer with Brisbane City Council and is responsible for the Council's activities with respect to Bushland Acquisition, Ecotourism and Fauna Management. Before joining the Council, Stephen was a freelance environmental consultant. This experience and his work with the Council have given him an in-depth knowledge of the ecology of the Greater Brisbane Region. His interests are widespread, but he is particularly concerned about koala habitat management and planning in Southeast Queensland. His position with the Council has dual roles in terms of interacting with local fauna and planning for their protection.

ENDNOTES

BEYOND THE BACKYARD

1. The Greater Brisbane Region, although not officially defined, is usually regarded as the Brisbane Statistical Division and a further 60-100 kilometres. It spreads west to the Great Dividing Range; north to around Nambour; and south towards the McPherson Range which forms the border between New South Wales and Queensland. For the purposes of this book, some areas outside this region are included because they are heavily visited by visitors to and residents of Brisbane.

2. Steel, J.G., 1975, *Brisbane Town in Convict Days 1824–1842*, Library Board of Queensland, Brisbane.

3. Horton, Helen, 1988, *Brisbane's Back Door — The Story of the D'Aguilar Range*, Boolarong Publications, Brisbane.

4. Davies, Wally, 1983, *Wildlife of the Brisbane Area*, Jacaranda, Brisbane.

5. Catterall, Carla, 1990, *A Natural Area Conservation Strategy for Brisbane City*, Brisbane City Council, Brisbane.

HABITATS

1. Churchett, Graham, 1982, *All in a Day's Walk — The Flora and Fauna of Lamington National Park*, Churchett, SA.

2. Catterall, 1990.

WILD PLACES — METROPOLITAN AREA

1. Low, Tim, 1995, Fauna of Brisbane, unpublished Brisbane City Council document.

2. Venables, Ian, 1996, Bird listing of Boondall Wetlands.

3. Bowden, J., 1993, Boondall Wetlands Advisory Committee, unpublished Brisbane City Council report.

4. Gynther, Dr I. and Caneris, A., 1995, *Summary of the Boondall Wetlands Reserve*, Nature Search survey.

5. Species list provided by John Bowden

6. Brisbane Conservation Atlas, 1983, unpublished Brisbane City Council.

7. Kordas, G., Coutts, R.H., Catterall, C.P., 1993, The Vegetation of Karawatha Forest and its significance in the South-east Queensland Landscape, unpublished report for Karawatha Protection Society Inc.

8. Stewart, D., 1995, Fauna Survey of Karawatha Forest, unpublished Brisbane City Council.

9. Dr Frank Carrick, personal communication.

10. Draft Tinchi Tamba Management Plan, 1994, unpublished Brisbane City Council.

11. Nudgee Beach Environmental Education Centre, personal communication.

12. Draft Tinchi Tamba Management Plan, 1994.

13. Coutts, Robert, 1983, "Toohey Forest" in *Wildlife of the Brisbane Area*, ed. Wally Davies, Jacaranda, Brisbane.

14. Toohey Forest Management Plan, 1993, unpublished Brisbane City Council document.

WILD PLACES — WITHIN 100 KM

1. Durrant B. and McRae I., 1994, *Birds of Pumicestone Passage*, BIEPA Inc. Bribie Island.

2. Henry, Don, 1983, "Moreton Island" in *Wildlife of the Brisbane Area*, ed. Wally Davies, Jacaranda, Brisbane.

3. Henry, 1983.

4. Heath, A.A. and others, 1976, *Moreton Island Environmental Impact Study and Strategic Plan*, Report to the Coordinator General's Department, Queensland.

5. Coleman, R., Covacevich, J. and Davie, P., 1984, *Focus on Stradbroke*, Boolarong Publications, Brisbane.

6. Guyatt, 1988, *A Natural History of Tamborine Mountain*, Tamborine Natural History Association.

7. Durbidge, E. and Covacevich J., 1981, *North Stradbroke Island*, Stradbroke Island Management Organisation.

INDEX OF COMMON NAMES

INDEX OF SCIENTIFIC NAMES

USEFUL CONTACTS

The information listed here is correct at time of printing. However, telephone numbers around Australia are being altered to incorporate an extra digit. This program will continue until 1999, please check numbers before dialling.

Queensland Museum Reference Centre
Ph: (07) 3840 7635
Fax: (07) 3846 1918

Australian Conservation Foundation
340 Gore St
FITZROY VIC 3065
Contact: Information Services
Ph: (03) 9416 1166 Fax: (03) 9416 0767

Australian Rainforest Conservation Society
19 Colorado Av
BARDON QLD 4065
Contact: Dr Aila Keto
Ph: (07) 3368 1318 Fax: (07) 3368 3938

Bayside Regional Park Advisory Committee
GPO Box 1434
BRISBANE QLD 4001
Contact: Peter Shilton/
Andrew Chamberlin
Ph: (07) 3403 6396 Fax: 3403 6413

Boondall Wetlands Management Committee
GPO Box 1434
BRISBANE QLD 4001
Contact: Peter Shilton/Fiona Chandler
Ph: (07) 3225 6396 Fax: (07) 3225 6413

Bribie Island Environmental Protection Association
PO Box 350
BRIBIE ISLAND QLD 4507
Contact: Shirley Elliot
Ph: (07) 3408 9286 Fax: (07) 3408 3577

BRISBANE CITY COUNCIL

Bushland Care Program
290 Lancaster Rd
ASCOT QLD 4007
Contact: Jenny Leask
Ph: (07) 3403 7173
Northside (07) 3403 7173
Southside (07) 3403 5948

Environment Management Branch
Floor 5
Brisbane Administration Centre
69 Ann Street
GPO Box 1434
BRISBANE QLD 4001

Brisbane River:
Annette Magee (07) 3225 6554
(07) 3403 6554

Bushland:
Ian Hislop (07) 3403 6776

Downfall Creek Bushland Centre:
(07) 3403 5937

Ecotourism:
Stephen Poole (07) 3403 3427

Environmental Education:
(07) 3403 6490

General Bushland Protection, Planning and Voluntary Conservation Agreements:
(07) 3403 3724

Fire Management:
Bryn Gullen (07) 3403 6369

Natural Area Reserve Management Planning:
Peter Shilton (07) 3403 6396

Open Space Planning:
Frank Andrews (07) 3403 3724
Joe Mumford (07) 3225 6778

Waterways:
Paul Mack (07) 3403 6882

Wetlands:
Patrick Bourke (07) 3403 4142
Hugh Suttor (07) 3403 9096

Vegetation Protection Orders:
Carole Rayner (07) 3403 6766

Parks and Gardens Branch
Floor 6
Brisbane Administration Centre
69 Ann Street
GPO Box 1434
BRISBANE QLD 4001

Natural Area Management:
General (07) 3403 9413

Boondall Wetlands Reserve:
Tinchi Tamba Wetlands Reserve,
Deagon Wetlands (07) 3403 7170

Karawatha Forest Park:
Toohey Forest Park (07) 3403 5948

Bayside Regional Park:
Brisbane Koala Park (07) 3403 5438

Mt Coot-tha Forest Park:
(07) 3403 2533

Brisbane Forest Park Administration Authority
60 Mt Nebo Road
THE GAP QLD 4061
Ph: (07) 3300 4855 (07) 3300 5347

Department of Environment *see*
National Parks and Wildlife Service

DEPARTMENT OF NATURAL RESOURCES
(State Forestry Service)
Moreton District Office
Floor 4 Mineral House
41 George Street
GPO BOX 2692 BRISBANE 4001
Ph: (07) 3224 2928 Fax: (07) 3224 2933

Downfall Creek Bushland Centre
815 Rode Road
MCDOWALL QLD 4053
Contact: Peter Armstrong
Ph: (07) 3403 5937
Fax: (07) 3353 2818

Environmental Defenders Office
(Qld) Inc.
2/133 George Street
BRISBANE QLD 4000
Contact: Jo Bragg
Ph: (07) 3210 0275 Fax: (07) 3210 0253

Forestry *see*
Department of Natural Resources

Greening Australia
GPO Box 9868
BRISBANE QLD 4001
Contact: The Secretary
Ph: (07) 3844 0211 Fax: (07) 3844 0727

Ipswich Envirocare Association Inc.
PO Box 2230
NORTH IPSWICH QLD 4305
Contact: Rocco de Pierri
Ph: (07) 3202 2906

Karawatha Forest Advisory
Committee
GPO Box 1434
BRISBANE QLD 4001
Contact: Peter Shilton/
Andrew Chamberlin
Ph: (07) 3403 6369 Fax: (07) 3403 6413

Karawatha Forest Protection Society
Contact: Bernie Volz
Ph: (07) 3209 4310

Landcare — South East Region
PO Box 96
IPSWICH 4305
Contact: Sam Brown
Ph: (07) 3280 Fax: (07) 3812 1715

Lota-Manly West Community Association Inc.
10 Boondarra Street
MANLY WEST QLD 4179
Contact: Ann Edwards
Ph: (07) 3393 5290

Macleay Island Conservation Group
Lot 27 Duncan Street
MACLEAY ISLAND QLD 4184
Contact: Leigh Abbot
Ph: (07) 3409 5642

Men of the Trees
PO Box 283
CLAYFIELD QLD 4011
Contact: Mrs Ngairetta Brennan
Ph: (07) 3262 1096

Moreton Island Protection Committee
PO Box 544
INDOOROOPILLY QLD 4068
Contact: Kay Martin
Ph: (07) 3378 0822

Mt Nebo and Mt Glorious
Environmental Protection Association
c/- Mt Nebo Post Office
MT NEBO QLD 4520
Contact: Peter Stevenson
Ph: (07) 3289 0223

National Parks Association of Queensland
PO Box 1040
MILTON QLD 4064
Contact: George Haddock
Ph: (07) 3367 0878 Fax: (07) 3367 0890

Queensland Conservation Council
PO Box 12046
BRISBANE
ELIZABETH STREET QLD 4000
Contact: Nicky Hungerford
Ph: (07) 3221 0188 Fax: (07) 3229 7992

QUEENSLAND NATIONAL PARKS AND WILDLIFE SERVICE

Central Moreton District (Greater Brisbane)
Wildlife Ranger
Qld National Parks and Wildlife Service
PO Box 42
KENMORE QLD 4069
Ph: (07) 3202 0200.

North Coast District
Wildlife Ranger
Qld National Parks and Wildlife Service
PO Box 168
COTTON TREE QLD 4558
Ph: (07) 4438 944.

South Coast District
Wildlife Ranger
Qld National Parks and Wildlife Service
PO Box 612
BURLEIGH HEADS QLD 4220
Ph: (07) 5535 3714

School of Environmental Studies
Griffith University
NATHAN QLD 4111
Ph: (07) 3875 7519 Fax: 3875 7459

Stradbroke Island Management Organisation
PO Box 8 Point Lookout
NORTH STRADBROKE ISLAND QLD 4183
Contact: Ellie Durbidge
Ph: (07) 3409 8115

Sunshine Coast Environment Council
PO Box 269
NAMBOUR QLD 4560
Contact: Jenny De Hayr
Ph: (074) 41 5747 (074) 41 7478

Tinchi Tamba Wetlands Advisory Committee
GPO Box 1434
BRISBANE QLD 4001
Contact: Peter Shilton/Fiona Chandler
Ph: (07) 3403 6396 Fax: (07) 3403 6413

Toohey Forest Management Committee
GPO Box 1434
BRISBANE QLD 4001
Contact: Peter Shilton/Ben McMullen
Ph: (07) 3225 6396 Fax: (07) 3225 6413

Toohey Forest Protection Society
Contact: Rob Simpson
Ph: (07) 3345 4527

WILDLIFE PRESERVATION SOCIETY OF QUEENSLAND
2/133 George St
BRISBANE QLD 4000
Contact: Pat Comben
Ph: (07) 3221 0194 Fax: (07) 3221 0701

Bayside
PO Box 427
CAPALABA QLD 4157
Contact: Simon Batais
Ph: (07) 3831 6199 (Pager: 59648)

Brisbane Valley
Lot 27 Willaura Drive
COOMINYA QLD 4311
Contact: Coral Rishworth
Ph: (074) 26 4742

Brisbane West
296 Ferguson Rd
NORMAN PARK QLD 4170
Contact: Karen Wright
Ph: (07) 3399 1933

Caboolture
PO Box 1415
CABOOLTURE QLD 4510
Contact: Eileen Rigdon
Ph: (074) 96 6644

Caloundra
PO Box 275
CALOUNDRA QLD 4551
Contact: Jill Chamberlain
Ph: (074) 442 707

East Logan
PO Box 1113
SPRINGWOOD QLD 4127
Contact: Ted Fensom
Ph: (07) 3341 6790

Gold Coast/Hinterland
PO Box 2441
SOUTHPORT QLD 4215
Contact: Elaine Quickenden
Ph: (0755) 33 1290

Pine Rivers
PO Box 377
STRATHPINE QLD 4500
Contact: Judy Elliott
Ph: (07) 3221 0194 Fax: (07) 3221 0701

Samford Valley
PO Box 272
SAMFORD VALLEY QLD 4520
Contact: Heather Holcroft
Ph: (07) 3289 7247 Fax: (07) 3289 7247

South Redlands
6 Kruger St
REDLAND BAY QLD 4165
Contact: Terri Guardala
Ph: (07) 3829 0809

Woogaroo Creek Community Group
2 Mill Street
GOODNA QLD 4300
Contact: Keith McCost
Ph: (07) 3288 4709

Brisbane City

Brisbane City Council, through its Environment Management Initiatives, is determined to protect the region's natural heritage and has put in place one of Australia's most comprehensive environment programs:

Bushland Acquisition

This targets natural areas of strategic importance. To date approximately $30 million has been spent acquiring 1300 hectares of "at risk" bushland. A major focus is the protection of koala habitat. However, land prices in Brisbane are high and only a small percentage of "at risk" bushland can be protected in this way.

Community Management Committees

Council establishes community-based advisory committees to coordinate management of acquisition areas. This brings local people into the decision-making process and facilitates community education, appreciation and participation in wildlife conservation.

Vegetation Protection Ordinances

VPOs are places on properties to protect native vegetation. They can be applied to single trees or to all vegetation on site. VPOs have decreased the rate of bushland clearance and rate rebates are given to affected land owners.

Zoning

A new town planning zone has been introduced with some zones having a specific conservation intent. The Conservation Zone will be progressively applied to bushland acquisition sites. Parts of the Non Urban Zone have also been redefined to provide for conservation objectives.

Voluntary Conservation Agreements

These are currently being developed between Council and some landowners to further protect against vegetation loss and to assist private landowners to better manage their bushland. Financial assistance will be used as an incentive to encourage participation.

Bushland Rehabilitation

A comprehensive program is under way across Brisbane to restore and rehabilitate remnant bushland. In addition to managing the major natural area reserves and other bushland parks (eg. with fencing, fire management, weed and rubbish removal and visitor facilities), approximately 120 ha of the most degraded bushland and waterway corridors have been intensely rehabilitated since 1991.